# Barron's Regents Exams and Answers

## Geometry

**Andre Castagna, Ph.D.**
Mathematics Teacher
Albany High School
Albany, New York

Barron's Educational Series, Inc.

## Dedication

To my loving wife Loretta, who helped make this endeavor possible with her unwavering support; my geometry buddy, Eva; and my future geometry buddies, Rose and Henry, and my professional colleagues for their patience and understanding.

*All inquiries should be addressed to:*
Barron's Educational Series, Inc.
250 Wireless Boulevard
Hauppauge, NY 11788
**www.barronseduc.com**

ISBN: 978-1-4380-0763-2
Library of Congress Catalog Card Number: 2015943380

PRINTED IN CANADA
9 8 7 6 5 4 3 2

# Contents

**Preface**                                                                v

**About the Exam**                                                         1

**Test-Taking Tips**                                                      13

**A Brief Review of Key Geometry Facts and Skills**                       21

  1. Angle, Line, and Plane Relationships                       21
  2. Triangle Relationships                                      39
  3. Constructions                                               51
  4. Transformations                                            65
  5. Triangle Congruence                                        85
  6. Coordinate Geometry                                        99
  7. Similar Figures                                           117
  8. Trigonometry                                              141
  9. Parallelograms                                            157
 10. Coordinate Geometry Proofs                                     173
 11. Circles                                                        191
 12. Solids                                                         213

**Glossary of Geometry Terms**                                          241

## Regents Examinations, Answers, and Self-Analysis Charts     253

June 2015 Exam......................................255
August 2015 Exam...................................313
June 2016 Exam......................................361
August 2016 Exam...................................417
June 2017 Exam......................................466
August 2017 Exam...................................516

## Appendix: The Common Core Geometry Learning Standards     565

# Preface

This book is designed to prepare you for the New York State Geometry Common Core Regents and is aligned to the Common Core Geometry Learning Standards. The key features you will find in this book are:

- **Test-Taking Tips** Advice and test-day strategies are provided to help you earn as many points as possible on the exam.

- **Brief Review of Geometry Concepts** A concise summary of the theorems, formulas, and math skills needed for success on the Regents exam is provided. Much of the information is presented in bulleted or table form to make it easy for you to use as a refresher and study aid. This summary is intended to supplement the more thorough presentation of high school geometry found in the Barron's *Lets Review: Geometry* book.

- **Glossary** A full glossary is found at the end of the Brief Review section. Be sure to look up any word that you are unfamiliar with as you work through the practice problems. Geometry is a vocabulary-intensive subject.

- **Actual Regents Exams and Step-by-Step Solutions with Explanations** Each of the problems in this book has a detailed solution. All steps in the solution are shown, along with an explanation of the relevant geometry theorems and concepts applied. Solutions to many of the multi-step problems begin with a summary of the big-picture strategy to be used in that problem. These solutions are an excellent way to strengthen your geometry knowledge. Understanding why particular strategies and concepts were chosen will make you a much more effective problem solver when you come across similar problems. Note that solving linear equations is an algebraic tool that is used frequently in Regents problems. It is expected that you have mastered that skill, and not all steps in solving linear equations are explicitly explained. For example, if you have the equation $2x + 4 = 12$, then the next line of the solution may read $2x = 8$. It is assumed that you understand that we subtracted 4 from each side in order to separate the variable from the constants in the equation.

- **Self-Analysis Charts** There is a self-analysis chart at the end of each Regents exam. The chart lists the main topic areas and shows which questions fall into which category. Use the chart to record how many points were earned on that practice exam, broken down by each topic area. You can then target areas of weakness for follow-up study and practice. You can also use the charts to find problems covering similar topics on other Regents exams in this book.

# About the Exam

## TEST LAYOUT AND STANDARDS

The common core geometry Regents is a 3-hour exam and consists of four parts:

| Part | Number of Questions | Type of Question | Points per Question | Total Number of Points |
|------|---------------------|------------------|---------------------|------------------------|
| I | 24 | Multiple-choice | 2 | 48 |
| II | 7 | Constructed-response | 2 | 14 |
| III | 3 | Constructed-response | 4 | 12 |
| IV | 2 | Constructed-response | 6 | 12 |
| Total | 36 | — | — | 86 |

Part I consists of 24 multiple-choice questions worth 2 points each. Part II consists of 7 constructed-response questions worth 2 points each, and part III consists of 3 constructed-response questions worth 4 points each. Part IV consists of 2 constructed-response questions worth 6 points each. One part IV question will either involve a multiple-step proof or ask you to develop an extended logical argument. The second question will require you to use modeling to solve a real-world problem.

The complete set of Geometry Common Core Learning Standards are listed in the Appendix. They are grouped in domains, and each domain accounts for a specified percentage of the total points on the exam, shown in the accompanying table. Note that some domains account for a far greater percentage of points than others. The domains are further divided into clusters with different levels of emphasis on the exam—major, supporting, and additional.

| Domain | Cluster | Emphasis | Standard[1] |
|---|---|---|---|
| Congruence (27–34%) | Experiment with transformations in the plane | Supporting | C-CO.1 through G-CO.5 |
| | Understand congruence in terms of rigid motions Prove geometric theorems | Major | G-CO.6 through G-CO.11 |
| | Make geometric constructions | Supporting | G-CO.12 G-CO.13 |
| Similarity and right triangle trigonometry (29–37%) | Understand similarity in terms of similarity transformations | Major | G-SRT.1 G-SRT.2 G-SRT.3 |
| | Prove theorems involving similarity | Major | G-SRT.4 G-SRT.5 |
| | Define trigonometric ratios and solve problems involving right triangles | Major | G-SRT.6 G-SRT.7 G-SRT.8 |
| Circles (2–8%) | Understand and apply theorems about circles | Additional | G-C.1 G-C.2 G-C.3 |
| | Find arc lengths and areas of sectors of circles | Additional | G-C.5 |

| Domain | Cluster | Emphasis | Standard[1] |
|---|---|---|---|
| Expressing geometric properties with equations (12–18%) | Translate between the geometric description and the equation for a conic section | Additional | G.GPE-1 |
| | Use coordinates to prove simple geometric theorems algebraically | Major | G-PE.4 through G-PE.7 |
| Geometric measurements and dimensions (2–8%) | Explain volume formulas and use them to solve problems | Additional | G-GMD.1 G-GMD.3 |
| | Visualize relationships between 2D and 3D objects | Additional | G-GMD.4 |
| Modeling with geometry (8–15%) | Apply geometric concepts in modeling situations | Major | G-MG.1 G-MG.2 G-MG.3 |

[1]See the Appendix for a complete list of the Common Core Geometry Learning Standards.

# FREQUENCY OF TOPICS

The actual distribution of questions in the last few Regents exams is shown in the following table.

| Topic | June 2015 | August 2015 | June 2016 | August 2016 | June 2017 | August 2017 |
|---|---|---|---|---|---|---|
| 1. *Angle and Segment Relationships* (define geometric terms and figures, vertical angles, complementary angles, supplementary angles, bisectors and midpoints, parallel and perpendicular lines, proofs of angle and segment relationships) | 17, 32 | | | 1, 11 | | |
| 2. *Angle and Segment Relationships in Triangles and Polygons* (angle sum theorem, exterior angle theorem, isosceles triangle theorem, equilateral triangles, interior and exterior angles of polygons, midsegments, centroids, central angles, proofs of triangle and polygon theorems) | 34 | 8, 11 | 9, 13 | 4, 8 | 10, 17 | 11, 16 |
| 3. *Constructions* (copy a segment, copy an angle; bisect a segment, bisect an angle; perpendicular bisector, perpendicular lines, parallel lines, equilateral triangle, square, regular hexagon, circle inscribed in a triangle, circle circumscribed about a triangle) | 25 | 26 | 31 | 28, 32 | 2 | 28 |
| 4. *Transformations* (definitions and properties of the rigid motions and dilations, preserved properties, compositions of transformations, mapping a figure onto itself with rotations) | 2, 4, 10, 16, 33 | 6, 7, 13, 20, 34 | 4, 8, 16, 25, 34 | 2, 9, 27, 29 | 1, 6, 7, 14, 22, 30, 31, 32 | 2, 6, 22, 27 |
| 5. *Transformations on the Coordinate Plane* (transformations of lines, transformation of points and figures) | 18, 22 | 5, 23, 24 | 3, 6, 12, 14, 22, 23, 26 | 5, 26, 33a | | |
| 6. *Triangle Congruence* (SAS, SSS, ASA, AAS, HL criteria; proofs involving triangle congruence; explain the congruence criteria in terms of rigid motions; prove congruence using rigid motions) | 24, 30 | 2, 30, 34 | 7, 35 | | 9, 33 | 30, 35 |
| 7. *Lines and Segments on the Coordinate Plane* (slope, distance, midpoint, dividing a segment in a specified ratio, equations of lines, determine if a point lies on curve given its equation, area, and perimeter) | 3, 9, 27 | 10, 31 | 3, 6, 12, 14, 22, 23, 26 | 14, 15, 16, 18, 21, 30 | 2, 12, 15, 19 | 3, 10, 17, 24, 29, 31 |

| Topic | June 2015 | August 2015 | June 2016 | August 2016 | June 2017 | August 2017 |
|---|---|---|---|---|---|---|
| 8. *Circles on the Coordinate Plane* (equations and graphs of circles, use completing the square to identify the center and radius of a circle) | 14 | 9 | 3, 6, 12, 14, 22, 23, 26 | | 2, 12, 15, 19 | 3, 10, 17, 24, 29, 31 |
| 9. *Similarity* (proportional corresponding parts, similar triangle proofs, scaled drawings, perimeter and area of similar figures, proportions in right triangles, prove the Pythagorean theorem) | 11, 15, 21, 31, 34 | 12, 14, 17, 19, 27, 29 | 2, 5, 17, 21, 27 | 10, 12, 22 | 5, 24, 29 | 5, 7, 9, 18 |
| 10. *Trigonometry* (finding a missing side of a right triangle, applications of the sine, cosine, and tangent in word problems, angle of elevation and depression, finding an angle using an inverse trigonometric function, cofunction relationships) | 5, 12, 28, 31 | 4, 32 | 11, 15, 28, 30 | 3, 6, 34 | 3, 13, 21, 36 | 15, 19, 21 |
| 11. *Parallelograms and Trapezoids* (classifying figures, properties of parallelograms and trapezoids, proofs involving parallelograms and trapezoids, transformations of parallelograms and trapezoids) | 13, 26, 33 | 1, 8, 28, 35 | 9, 33 | 7, 24 | 11, 20 | 8, 14, 26 |
| 12. *Coordinate Geometry Proofs* (proofs on the coordinate plane) | 36 | 22, 33 | | 33b | 35 | 32 |
| 13. *Circles* (similarity of circles, central and inscribed angles, angles and segments formed by chords, tangents, and secants, quadrilaterals inscribed in a circle, radian measure, arc length, area of a sector) | 8, 20, 29 | 11, 12, 15, 18 | 10, 24, 29 | 19, 23, 25, 35 | 4, 8, 26 | 4, 12, 23, 33 |
| 14. *Volume* (prisms, cylinders, cones, pyramids, spheres, properties of solids, calculating volume of solids, justify the formulas for circumference and area of a circle, justify the volume formulas, Cavalieri's principle, identify cross sections and solids of revolution) | 1, 6, 23 | 3, 21 | 1, 20, 36 | 3, 13, 20 | 16, 18, 27, 28 | 1, 13, 25 |
| 15. *Modeling* (modeling real-world situations with solids and planar figures, applications of density, design problems involving area and volume) | 7, 19, 35 | 16, 25, 36 | 18, 32 | 17, 36 | 23, 34 | 20, 34, 36 |

# HOW THE TEST IS SCORED

All multiple-choice questions are worth 2 points each, and no partial credit is awarded. The constructed-response questions are worth 2, 4, or 6 points each. These are graded according to the scoring rubric issued by the New York State Department of Education. The rubric allows for partial credit if an answer is partially correct.

When determining partial credit, the rubrics often distinguish between *computational* errors and *conceptual* errors. A computational error might be an error in algebra, graphing, or rounding. Computational errors generally will cost you 1 point, regardless of whether the question is a 2-point or 6-point question. Conceptual errors include using the incorrect formula (volume of a cone instead of a prism) or applying an incorrect relationship (congruent instead of supplementary same side interior angles). Half the credit of the problem is generally deducted for conceptual errors.

Because partial credit is awarded on the constructed-response questions, it is extremely important to show all your work. A correct answer with no work shown will usually cost you most of the points available for that problem.

Your total number of points earned for all four parts will be added to determine your raw score. A conversion table specific for each exam will then be used to convert your raw score to a final scaled score, which is reported to your school. The actual conversion charts are included at the end of each of the Regents exams in this book so you can convert your raw score to a final scaled score. The percentage of points needed to earn a final score of 65% is usually less than 65%, but the effect of the "curve" diminishes as your raw score increases.

# CALCULATOR, COMPASS, STRAIGHTEDGE, PEN, AND PENCIL

Graphing calculators are required for the Geometry Common Core exam, and schools must provide one to any students who don't have their own. You may bring your own calculator if you own one, and many students feel more comfortable using their own familiar calculator. Be aware, though, that test administrators may clear the memory and any stored programs you might have saved on your calculator before the exam begins.

Any calculator provided to you will likely have had its memory cleared as well. This process will restore the calculator to its default settings, which may not be the familiar ones you are accustomed to using. The most common setting that may affect your work is the degree versus radian mode. Some calculators have radian mode as the default. It is a good idea to find out what calculator will be provided to you and how to switch between degree and radian mode.

A good working knowledge of the graphing calculator will let you apply more than one method to solving a problem. Some calculator techniques worth knowing are

- Using the graph and intersect feature to solve linear and quadratic equations
- Using tables to find patterns or to quickly apply trial and error to solve a problem
- Finding the square root and cube root of a number
- Using the trigonometric and inverse trigonometric functions

You will also be provided with a compass and straightedge if you do not have your own. Bringing your own compass is highly recommended, since the compasses provided by your school are of unknown quality. Also, it will be less stressful on test day to work with a compass that you are familiar with.

The straightedge provided to you should be used only for making straight lines when graphing or doing constructions. The straightedge may have length marking in inches or centimeters, but these should not be used for determining the length of a segment or checking if two segments are congruent.

You are responsible for bringing your own pen and pencil to the exam. All work must be done in blue or black pen, with the exception of graphs and diagrams which may be done in pencil.

## SCRAP PAPER AND GRAPH PAPER

You are not allowed to bring your own scrap paper or graph paper into the exam. The exam booklet contains one page of scrap paper and one page of graph paper. These are perforated and can be removed from the booklet to make working with them easier. Any work you put on these sheets is *not* graded. If you want a grader to consider any work there, you must copy it into the appropriate space in the booklet.

## REFERENCE SHEET

The following reference sheet is provided to you during the Regents exam. The same reference sheet is used for the Algebra I, Geometry, and Algebra II exams. The three sequence and series formulas and the exponential growth and decay formulas are not part of the Common Core Geometry curriculum, so do not be concerned with them. You should be familiar with all the other formulas. Remember, anytime you are asked to calculate an area or volume, check the reference sheet. Many of those formulas are provided.

# COMMON CORE HIGH SCHOOL MATH REFERENCE SHEET

(Algebra I, Geometry, Algebra II)

## CONVERSIONS

| | |
|---|---|
| 1 inch = 2.54 centimeters | 1 ton = 2000 pounds |
| 1 meter = 39.37 inches | 1 cup = 8 fluid ounces |
| 1 mile – 5280 feet | 1 pint = 2 cups |
| 1 mile = 1760 yards | 1 quart = 2 pints |
| 1 mile = 1.609 kilometers | 1 gallon = 4 quarts |
| 1 kilometer = 0.62 mile | 1 gallon = 3.785 liters |
| 1 pound = 16 ounces | 1 liter = 0.264 gallon |
| 1 pound = 0.454 kilogram | 1 liter = 1000 cubic centimeters |
| 1 kilogram = 2.2 pounds | |

## FORMULAS

| | |
|---|---|
| Triangle | $A = \frac{1}{2}bh$ |
| Parallelogram | $A = bh$ |
| Circle | $A = \pi r^2$ |
| Circle | $C = \pi d$ or $C = 2\pi r$ |
| General Prisms | $V = Bh$ |
| Cylinder | $V = \pi r^2 h$ |
| Sphere | $V = \frac{4}{3}\pi r^3$ |
| Cone | $V = \frac{1}{3}\pi r^2 h$ |

| Pyramid | $V = \frac{1}{3}Bh$ |
|---|---|
| Pythagorean Theorem | $a^2 + b^2 = c^2$ |
| Quadratic Formula | $x = \frac{-b \pm \sqrt{b^2 - 4ac}}{2a}$ |
| Arithmetic Sequence | $a_n = a_1 + (n - 1)d$ |
| Geometric Sequence | $a_n = a_1 r^{n-1}$ |
| Geometric Series | $S_n = \frac{a_1 - a_1 r^n}{1 - r}$ where $r \neq 1$ |
| Radians | $1$ radian $= \frac{180}{\pi}$ degrees |
| Degrees | $1$ degree $= \frac{\pi}{180}$ radians |
| Exponential Growth/Decay | $A = A_0 e^{k(t - t_0)} + B_0$ |

# PREPARING FOR THE EXAM

- Preparing to earn a high grade on the Geometry Regents begins well before exam day. Don't try to cram for the Geometry Regents the day before the exam, or even the week before. Successful students begin preparing months ahead of time!
- Make the most of your class time. Take good notes, attempt all homework and classwork, and—most importantly—ask questions when you encounter something you don't understand. Check and correct your work, especially exams and quizzes. Save your exams and quizzes in a folder or binder, they are an excellent resource for review and can help you document your progress.
- You should be working with a good review book throughout the school year. The recommended review book is Barron's *Let's Review: Geometry*. You will be able to start reviewing with that book topic by topic before you reach the end of the curriculum in your geometry class.
- Use the self-analysis charts at the end of each Regents exam in this book to record the number of points you earn within each topic area, and use those results to help identify areas for additional practice.
- Work through at least one Regents exam under "exam conditions"—no distractions, no breaks, and a 3-hour time limit. Measure how much time you spend on each part, and use that as a guide to help establish a time-management plan for the actual exam.
- Review any vocabulary words you are unfamiliar with. Geometry is a vocabulary-intensive subject, and understanding the full meaning and implication of a word encountered in a problem is critical to being able to solve the problem. The glossary at the end of this book is a good resource. One good strategy is to make flash cards with a word on one side and figure on the other.
- Work through as many practice problems as you can. A greater variety of problems that you encounter beforehand means a greater probability of seeing familiar problems on the Regents.

- Become familiar with the formula sheet. You should know what is on the formula sheet and what's not. Also, you should know what each of the symbols and variables represents.
- Read through the step-by-step solutions to the Regents questions for any question you do not get right while practicing. Try to identify why you were not able to correctly answer the question. You will learn from your errors only if you correct them as soon as possible. Be aware of why the errors occur. Common errors are

    - Not remembering a formula or relationship
    - Having difficulty applying multiple steps or concepts in a single problem
    - Having weak algebra skills
    - Having a weak vocabulary
    - Not reading the question carefully
    - Not checking your work

# Test-Taking Tips

| TIP 1 |
| --- |
| **Be mentally prepared for the exam** |

**SUGGESTIONS**
1. Don't try to cram the night before the exam—prepare ahead of time.
2. Walk into the exam room with confidence.
3. You can retake the exam if the grade you earn does not meet the expectation you set for yourself, so don't panic.

---

### TIP 2

**Be physically prepared for the exam**

## SUGGESTIONS

1. Get plenty of sleep the night before, and set an alarm if necessary so you wake on time.
2. Eat a good breakfast the day of the test.
3. Dress in comfortable clothes.
4. Give yourself plenty of time to get to the test. Late students may be admitted up to a certain time, but they will not be given any additional time at the end of the exam. Check with your teachers for the current policies.

---

### TIP 3

**Come prepared with all materials**

## SUGGESTIONS

1. Be sure to have
   - Any identification required by your school
   - Time and room number of your exam
   - Two pens, two pencils, eraser, compass, ruler, and graphing calculator (if you have one)
   - Do **not** bring any restricted items such as cell phones, music or recording devices, food, and drinks. Ask your teacher ahead of time what items are prohibited. If you do need to bring a cell phone or other electronic device to school, plan to leave it in your locker or other secure place. Using a prohibited device during the exam will invalidate your score.

---

**TIP 4**

**Have a consistent plan for working through each problem**

---

## SUGGESTIONS

1. Start by reading the question carefully. Common errors to watch out for are

   - Not writing your answer in the correct form, such as a simplified radical or rounded decimal
   - Using the wrong method, such as graphical versus algebraic
   - Solving for a variable but not substituting to find a requested angle or segment measure
   - Missing key words that provide necessary information (parallel, isosceles, regular, "not necessarily true," etc.)

2. Sketch a figure if none is provided. Mark up the figures with relevant dimensions, congruence markings, parallel markings, and so on.

3. Try to break down complex multi-step problems into smaller pieces, looking for relationships that you can apply.

4. For proofs, decide if you are going to write a two-column or paragraph proof.

5. If you don't see how to get to a final answer at first, try applying any formulas or relationships that you recognize. Sometimes finding a certain length or angle may help you see the path to the final answer. Don't panic if you cannot immediately answer a question!

---

## TIP 5

### Budget your time

---

## SUGGESTIONS

1. Don't spend too much time on any one problem. If you are having difficulty with a problem, circle it and come back to it at the end. There are going to be a mix of easy, moderate, and difficult questions. You don't want to run out of time before doing all the easy problems.

2. Before the test day, record the time it takes you to do each part of one of the practice Regents exams in this book. Your goal should be to finish all parts in $2\frac{1}{2}$ hours, which will give you 30 minutes to check your work.

---

## TIP 6

### Express your answer in the form requested

---

## SUGGESTIONS

1. When required to round to a specified place value, do not round until you get your final answer.

2. When using formulas that involve $\pi$, check if the answer is to be "expressed in terms of $\pi$" or "rounded" to a certain place value. Expressed in terms of $\pi$ means leave $\pi$ out of the calculation, and then put in the $\pi$ when writing your final answer, for example "$32\pi$." If the answer is to be rounded, then you should type $\pi$ into your calculator along with the appropriate formula. Remember, always use the $\pi$ button, not 3.14 or any other approximation of $\pi$.

3. Simplify radicals when necessary.
4. If your answer to a multiple-choice question is not one of the choices, check to see if it can be rewritten as one of the choices by simplifying a fraction or radical or rearranging an equation. For example, the equation of a line may be written in multiple forms, or an answer may be expressed in terms of $\pi$ instead of a rounded equivalent.

---

**TIP 7**

**Maximize your partial credit**

---

## SUGGESTIONS

1. Show all work for parts II, III, and IV. Most of the credit is lost if you write a correct answer with no supporting work.
2. Write any formulas used in parts II, III, and IV, and then show each step of the evaluation.
3. Work through the entire problem even if you are not sure if you are correct! If you make an error along the way, but the following work is consistent with the error, you will likely receive partial credit for the correct method.
4. Do not leave any questions blank. There is no penalty for guessing.

---

## TIP 8

### Consider alternative methods to solving a problem

## SUGGESTIONS

1. Trial and error is a valid method for solving an equation or problem. However, you must clearly write the equation you are using along with at least *three* guesses with an appropriate check. Even if your first guess is correct, demonstrate that you understand the method by showing work for two incorrect guesses and the checks.

2. Some problems can be solved graphically even though a coordinate plane grid is not provided with the problem. You can use the provided scrap graph paper in these cases. The scrap graph paper is not graded, but this is not a concern for multiple-choice problems since work does not need to be shown. For some constructed-response questions, sketching the problem and solution on the graph paper may be a good way to check your result.

---

## TIP 9

### Eliminate obviously incorrect choices on multiple-choice questions

## SUGGESTIONS

1. You can sometimes rule out choices if you

   - Expect an obtuse or acute angle for an answer
   - Expect a length to be longer or shorter than another known length

2. Use the fact that unless otherwise specified, figures are approximately to scale

---

**TIP 10**

**Make your paper easy to grade**

---

## SUGGESTIONS

1. Work neatly. Your work should be legible and flow from top to bottom.
2. Cross out any incorrect work you do not want the graders to consider on constructed-response questions. You will lose credit if there is both a correct and incorrect answer.
3. Follow directions if you need to change an answer on a multiple-choice bubble sheet. Ask a proctor if you are unsure.

---

**TIP 11**

**Check your work**

---

## SUGGESTIONS

1. If you finish the exam with time left, go back and check as much of your work as you can. After taking the time to prepare for the exam, it would be a waste of that effort to not make use of any extra time you might have.

# A Brief Review of Key Geometry Facts and Skills

## 3.1. ANGLE, LINE, AND PLANE RELATIONSHIPS

### NOTATION

Points, segments, lines, and planes are represented using the following notation

| Name | Definition | Figure | Notation |
|------|-----------|--------|----------|
| Point | A location in space with no length, width, or thickness | $\bullet$ A | Point $A$ |
| Line | An infinitely long set of points that has no width or thickness | $\overset{\leftrightarrow}{A \quad B}$ <br><br> $\overset{\leftrightarrow}{M \quad N \quad O}$ <br><br> $\overset{\leftrightarrow}{\qquad}$ $m$ | $\overleftrightarrow{AB}$ <br><br> $\overleftrightarrow{MN}, \overleftrightarrow{NO},$ or $\overleftrightarrow{MO}$ <br><br> Line $m$ |
| Ray | A portion of a line with one endpoint and including all points on one side of the endpoint | $A \quad B \nearrow$ | $\overrightarrow{AB}$ |

| Name | Definition | Figure | Notation |
|---|---|---|---|
| Segment | A portion of a line bounded by two endpoints | | $\overline{FG}$ |
| Length of a segment | The distance between the two endpoints of a segment | | $\overline{FG}$ |
| Plane | A flat set of points with no thickness that extends infinitely in all directions | | Plane *ABC* <br><br> Plane *R* |
| Angle | A figure formed by two rays with a common endpoint | | ∠*RST* or ∠*S* |
| Angle measure | The amount of opening of an angle, measured in degrees or radians | | ∠1 <br> m∠1 |

An angle can be classified as a

- **Right angle**—measures 90°
- **Acute angle**—measures less than 90°
- **Obtuse angle**—measures greater than 90° and less than 180°
- **Straight angle**—measures 180°

## INTERSECTING, PARALLEL, PERPENDICULAR, AND SKEW

- The intersection of two lines is one point (lines $r$ and $s$ intersecting at point $M$).
- The intersection of two planes is one line (planes $ABC$ and $ABD$ intersect along $\overleftrightarrow{AB}$).

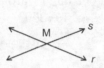
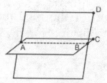

- *Coplanar lines* lie in the same plane, they are either parallel ($\overleftrightarrow{AD}$ and $\overleftrightarrow{BC}$) or intersect ($\overleftrightarrow{AD}$ and $\overleftrightarrow{AB}$).
- *Parallel lines* are coplanar and never intersect. The symbol for parallel is // ($\overleftrightarrow{AD}$ // $\overleftrightarrow{BC}$).
- *Skew lines* are lines that are not coplanar. $\overleftrightarrow{AE}$ and $\overleftrightarrow{GF}$ are skew.
- *Perpendicular lines* intersect at right angles, which measure 90°. The symbol for perpendicular is ⊥ ($r \perp s$). Line $r$ ⊥ line $s$ in the accompanying figure. The small square at the point of intersection is a symbol for a right angle. m∠$ABC$ + m∠$DEF$ = 180°.

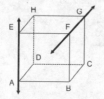

## BASIC ANGLE RELATIONSHIPS

- *Supplementary* angles have measures that sum to 180°. ∠ABC and ∠DEF are supplementary.

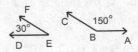

- *Complementary* angles have measures that sum to 90°. ∠1 and ∠2 are complementary. m∠1 + m∠2 = 90°.

- The sum of the measures of adjacent angles around a point equals 360°. m∠1 + m∠2 + m∠3 + m∠4 = 360°.

- The sum of the measures of adjacent angles around a line equals 180°. Two adjacent angles that form a straight line are called a *linear pair*, and are supplementary. m∠1 + m∠2 = 180°. ∠1 and ∠2 are a linear pair.

- The measure of a whole equals the sum of the measures of its parts. m∠ABC = m∠1 + m∠2 + m∠3.

- *Vertical angles* are the congruent opposite angles formed by intersecting lines. ∠1 and ∠3 are vertical angles; therefore, ∠1 ≅ ∠3. ∠2 and ∠4 are vertical angles; therefore, ∠2 ≅ ∠4.

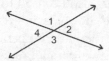

- *Angle bisectors* divide angles into two congruent angles. $\overrightarrow{CD}$ bisects ∠ACB, ∠ACD ≅ ∠BCD.

## ANGLES FORMED BY PARALLEL LINES

When two parallel lines (lines *m* and *n*) are intersected by a transversal (line *v*), eight angles are formed. Any pair of these angles will be either supplementary or congruent. Certain pairs are given special names. The named pairs have the following properties:

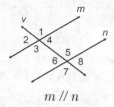

*m // n*

*Alternate interior angles are congruent*. Alternate interior angles lie between parallel lines *m* and *n* on opposite sides of transversal *v*.

*Same side interior angles are supplementary*. Same side interior angles lie between parallel lines *m* and *n* on the same side of transversal *v*. ∠4 and ∠5 are supplementary, ∠3 and ∠6 are supplementary. m∠4 + m∠5 = 180°, m∠3 + m∠6 = 180°.

*Corresponding angles are congruent.* Corresponding angles are the interior/exterior pair of congruent angles on the same side of the transversal. $\angle 4 \cong \angle 8$, $\angle 1 \cong \angle 5$, $\angle 2 \cong \angle 6$, $\angle 3 \cong \angle 7$.

These relationships can also be used to prove two lines are parallel. For example, if the alternate interior angles formed by two lines are congruent, then the lines are parallel.

If a problem involves two parallel lines, but no transversal that intersects both lines, it may be helpful to add an additional line called an auxiliary line. In order to find m$\angle BED$ in the accompanying figure, the auxiliary line $\overleftrightarrow{FEG}$ is drawn parallel to $\overline{CD}$ and $\overline{AB}$. Now the two alternate interior angle relationships can be used:

| | |
|---|---|
| m$\angle FED$ = m$\angle CDE$ = 38° | alternate interior angles |
| m$\angle FEB$ = m$\angle ABE$ = 32° | alternate interior angles |
| m$\angle DEB$ = m$\angle FED$ + m$\angle FEB$ | angle addition |
| = 38° + 32° | |
| = 70° | |

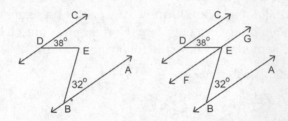

## BASIC LINE/SEGMENT/RAY RELATIONSHIPS

In segments, the point that divides a segment into two congruent segments is called a *midpoint*. In the accompanying figure, S is the midpoint of $\overline{RT}$, and $\overline{RS} \cong \overline{ST}$. Any line, segment, or ray that intersects a segment at its midpoint is called a *segment bisector*, or simply *bisector*. Line *m* bisects $\overline{RT}$ at its midpoint S. Lines can bisect segments, but lines cannot be bisected themselves because they do not have a finite length.

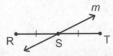

- A line, segment, or ray that is perpendicular to and bisects a segment is called a perpendicular bisector. $\overleftrightarrow{CD}$ is the perpendicular bisector of $\overline{AB}$. The four angles formed by their intersection are right angles, and $\overline{AE} \cong \overline{BE}$.

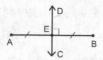

As with angles, the length of a divided segment is equal to the sum of its parts. In the accompanying figure, $MN + NO = MO$. If point $N$ is also a midpoint, then the additional relationship $MN - NO$ would be true.

## ANGLES IN POLYGONS
Definitions:

- **Polygon**—a closed planar figure with straight sides
- **Regular polygon**—a polygon whose sides and angles are all congruent
- **Interior angle**—the angle inside the polygon formed by two adjacent sides
- **Exterior angle**—the angle formed by a side and the extension of an adjacent side in a polygon
- **Triangle**—a polygon with 3 sides
- **Quadrilateral**—a polygon with 4 sides
- **Pentagon, hexagon, octagon, decagon**—polygons with 5, 6, 8, and 10 sides

In the accompanying pentagon, angles 1 through 5 are interior angles and angles 6 through 10 are exterior angles.

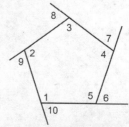

The following relationships can be used to find interior and exterior angles of a polygon:

- The sum of the measures of interior angles of a polygon with $n$ sides equals $180° \times (n-2)$.
- The measure of one interior angle of a regular polygon with $n$ sides equals $\dfrac{180°(n-2)}{n}$.
- The sum of the measures of exterior angles of a polygon with $n$ sides equals $360°$.
- The measure of one exterior angle of a regular polygon with $n$ sides equals $\dfrac{360°}{n}$.

Using the accompanying figure of the pentagon on page 27:

$$m\angle 1 + m\angle 2 + m\angle 3 + m\angle 4 + m\angle 5 = 180 \times (n-2).$$
$$= 180(5-2)$$
$$= 540°$$

$$m\angle 6 + m\angle 7 + m\angle 8 + m\angle 9 + m\angle 10 = 360°$$

If the pentagon is regular,

$$m\angle 1 = \frac{180°(n-2)}{n}$$
$$= \frac{180°(5-2)}{5}$$
$$= 108°$$

$$m\angle 6 = \frac{360°}{n}$$
$$= \frac{360°}{5}$$
$$= 72°$$

- A central angle is the angle formed by connecting the center of the polygon to two adjacent vertices, as shown. The measure of a central angle in a regular polygon with $n$ sides equals $\dfrac{360°}{n}$. For pentagon $ABCDE$, m$\angle APB = \dfrac{360°}{5} = 72°$.

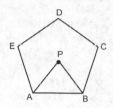

# Practice Exercises

1.  In the figure of the rectangular prism, which of the following is true?

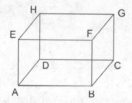

(1) Points $E$, $H$, $D$, and $A$ are coplanar and collinear.

(2) $\overline{HD}$ is skew to $\overline{CD}$, and $\overline{CD} \perp \overline{CG}$.

(3) $\overline{EA}$ // $\overline{CG}$, and $\overline{EH}$ is skew to $\overline{FB}$.

(4) $\overline{EA} \perp \overline{BC}$, and $\overline{AB}$ // $\overline{CD}$.

2.  Which of the following would be the best definition of an angle?

(1) the union of two rays with a common endpoint

(2) a geometric figure measured in degrees

(3) one-third of a triangle

(4) a line that bends

3. $\overrightarrow{BFA} \perp \overrightarrow{CF}$ and m$\angle CFE = 42°$. Find m$\angle BFD$.

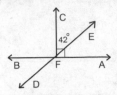

   (1) 42°          (3) 48°

   (2) 45°          (4) 52°

4. $\overrightarrow{BC}$ bisects $\angle ABD$. If m$\angle ABD = (8x - 12)°$ and m$\angle ABC = (3x + 4)°$, find m$\angle ABD$.

   (1) 10°          (3) 34°

   (2) 20°          (4) 68°

5. Line $p$ intersects lines $m$ and $n$. For what values of $x$ could you conclude that $m$ is parallel to $n$?

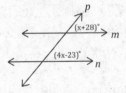

   (1) 12°          (3) 35°

   (2) 17°          (4) 62°

6. Points $F$, $G$, $H$, and $I$ are collinear. $G$ is the midpoint of $\overline{FH}$, $H$ is the midpoint of $\overline{GI}$, $\overline{FG} = (x^2 + 5x)$, and $\overline{HI} = 3x + 8$. What is the length of $\overline{FG}$?

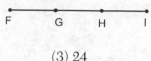

(1) 14    (3) 24
(2) 20    (4) 32

7. In the accompanying figure, lines $m$ and $n$ are parallel. If $m\angle1 = 32°$ and $m\angle2 = 78°$, what is the measure of $\angle3$?

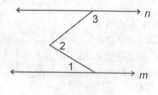

(1) 102°    (3) 134°
(2) 110°    (4) 148°

8. A regular polygon has interior angles that each measure 156°. How many sides does the polygon have?

9. Regular octagon $ABCDEFGH$ is shown in the figure below.

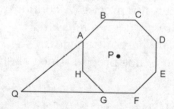

Sides $\overline{AB}$ and $\overline{FG}$ are both extended to $Q$.
What is the measure of $\angle Q$?
(1) 22.5°    (3) 45°
(2) 30°    (4) 60°

**10.** In regular hexagon $ABCDEF$, $\overline{AD}$ and $\overline{FC}$ intersect at $O$. What is the measure of $\angle AOF$?

**11.** In the accompanying figure, $\overline{WPX}$ intersects $\overline{YPZ}$ at $P$.

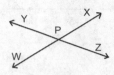

Prove the theorem that the opposite angles formed by intersecting lines are congruent, $\angle WPY \cong \angle ZPX$.

**Solutions**

  **1.** Choice (1) is not correct since $E$, $H$, and $D$ do not lie on a single straight line and are not collinear.

  Choice (2) is not correct since $\overline{HD}$ and $\overline{CD}$ are perpendicular, so they cannot be skew.

  Choice (4) is not correct since $\overline{EA}$ and $\overline{BC}$ do not intersect, so they cannot be perpendicular.

  The correct choice is (**3**).

  **2.** Considering each choice:

  (1) This definition is precise and uses previously defined geometric terms.

  (2) This definition is not precise. A counterexample would be an arc, which can also have a degree measure.

  (3) Not all angles are part of triangles.

  (4) This definition is not precise, since "bend" is not a defined geometric term.

  The correct choice is (**1**).

**3.** $m\angle CFE + m\angle EFA = 90°$      Perpendicular lines form
90° angles.

$$42° + m\angle EFA = 90°$$
$$m\angle EFA = 48°$$
$$m\angle BFD = m\angle EFA \quad \text{Vertical angles are congruent.}$$
$$m\angle BFD = 48°$$

The correct choice is **(3)**.

**4.** $m\angle ABD = m\angle ABC + m\angle CBD$      angle addition
$m\angle ABC = m\angle CBD$      bisector forms
     $2 \cong$ segments
$m\angle ABC + m\angle ABC = m\angle ABD$      substitute $m\angle ABC$ for
     $m\angle CBD$

$$3x + 4 + 3x + 4 = 8x - 12$$
$$6x + 8 = 8x - 12$$
$$x = 10$$

$$m\angle ABD = (8 \cdot 10 - 12)° \quad \text{evaluate } m\angle ABD \text{ for}$$
$$x = 10$$

$$= 68°$$

The correct choice is **(4)**.

**5.** The two angles indicated in the figure are corresponding angles, so lines $m$ and $n$ are parallel if the corresponding angles are congruent, or have the same angle measure.

$$4x - 23 = x + 28$$
$$3x = 51$$
$$x = 17$$

The correct choice is **(2)**.

**6.** A bisector divides a segment into two congruent halves, so $FG = GH$ and $GH = HI$. Combine these relationships to obtain $FG = HI$.

$$x^2 + 5x = 3x + 8$$
$$x^2 + 2x - 8 = 0$$
$$(x - 2)(x + 4) = 0 \qquad \text{factor}$$
$$(x - 2) = 0 \qquad (x + 4) = 0 \qquad \text{zero product property}$$
$$x = 2 \qquad x = -4$$

Eliminate the solution $x = -4$, because that would lead to a negative or zero length; therefore, $x = 2$.

$$FG = x^2 + 5x$$
$$= 2^2 + 5 \cdot 2$$
$$= 14$$

The correct choice is **(1)**.

**7.** Sketch the auxiliary line parallel to $m$ and $n$ as shown.

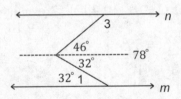

The 78° angle is divided into a lower and upper part. The lower part is an alternate interior angle to $\angle 1$, so it must measure 32°. The measure of the upper part is equal to $78° - 32° = 46°$.

$\angle 3$ and the 46° angle are same side interior angles, so they must be supplementary.

$$m\angle 3 + 46° = 180°$$
$$m\angle 3 = 134°$$

The correct choice is **(3)**.

**8.** Each interior angle of a polygon measures $\dfrac{180°(n-2)}{n}$ degrees.

$$\frac{180°(n-2)}{n} = 156$$
$$180(n - 2) = 156n$$
$$180n - 360 = 156n$$
$$24n = 360$$
$$n = 15$$

The polygon has 15 sides.

**9.** The sum of the exterior angles of a polygon equals 360°, and in a regular polygon each exterior angle measures $\dfrac{360°}{n}$.

The strategy is to find each interior angle of quadrilateral $QAHC$.

$$m\angle QAH = m\angle QGH = \frac{360°}{8}$$
$$= 45°$$

Each interior angle of a regular polygon is the supplement of an exterior angle.

$$m\angle AHG = 180° - 45° = 135°$$

The exterior convex angle $\angle AHG = 360° - 135° = 225°$.
The sum of the measures of the interior angles of

$$QAHG = (n - 2) \cdot 180°$$
$$= (4 - 2) \cdot 180°$$
$$= 360°$$

$$m\angle Q + m\angle QAH + m\angle AHG + m\angle QGH = 360°$$
$$m\angle Q + 45° + 225° + 45° = 360°$$
$$m\angle Q + 315° = 360°$$
$$m\angle Q = 45°$$

The correct choice is **(3)**.

**10.** The intersection of the two diagonals $\overline{FC}$ and $\overline{AD}$ is the center of the regular polygon, so $\angle AOF$ is a central angle.

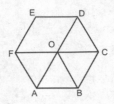

$\mathrm{m}\angle AOF = \dfrac{360°}{n}$     central angle formula for a regular polygon

$\qquad\quad = \dfrac{360°}{6}$

$\qquad\quad = 60°$

**11.** The strategy is to show $\angle WPY$ and $\angle ZPX$ are supplementary to the same angle.

| Statement | Reason |
|---|---|
| 1. $\overline{WPX}$ intersects $\overline{YPZ}$ at $P$. | 1. Given |
| 2. $\angle WPY$ and $\angle YPX$ are a linear pair.<br>$\angle ZPX$ and $\angle YPX$ are a linear pair. | 2. Definition of a linear pair |
| 3. $\angle WPY$ and $\angle YPX$ are supplementary.<br>$\angle ZPX$ and $\angle YPX$ are supplementary. | 3. Linear pairs are supplementary |
| 4. $\angle WPY \cong \angle ZPX$ | 4. Angles supplementary to the same angle are congruent |

## 3.2. TRIANGLE RELATIONSHIPS

### CLASSIFYING TRIANGLES

A triangle is a polygon with three sides.

Triangles can be classified by their sides or by their angles.

- By Sides

    **Scalene**—no sides are congruent.
    **Isosceles**—two sides are congruent.
    **Equilateral**—three sides are congruent.

- By Angle

    **Acute**—all angles are acute.
    **Right**—one angle is a right angle.
    **Obtuse**—one angle is obtuse.

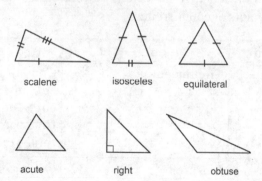

scalene  isosceles  equilateral

acute  right  obtuse

### SPECIAL SEGMENTS AND POINTS OF CONCURRENCY IN TRIANGLES

There are four special segments that can be drawn in a triangle, and every triangle has three of each. These are the altitude, median, angle bisector, and perpendicular bisector, shown in the accompanying figure.

*Altitude*—a segment from a vertex perpendicular to the opposite side

*Median*—a segment from a vertex to the midpoint of the opposite side

*Angle bisector*—a line, segment, or ray passing through the vertex of a triangle and bisecting that angle

*Perpendicular bisector*—a line, segment, or ray that is perpendicular to and passes through the midpoint of a side

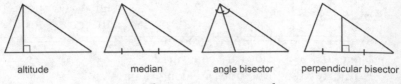

| altitude | median | angle bisector | perpendicular bisector |

Special segments in triangles

The three medians of a triangle will all intersect at a single point called the centroid. We say the medians are *concurrent* at the centroid, or the centroid is the *point of concurrency*. The other special segments will be concurrent as well. The following table illustrates each of the points of concurrency.

| Segment | Concurrent at | Feature |
|---------|---------------|---------|
| Perpendicular bisectors | Circumcenter | Center of the circumscribed circle, equidistant from each of the three vertices |
| Angle bisectors | Incenter | Center of the inscribed circle, equidistant from the three sides of the triangle |
| Medians | Centroid | Divides each median in 2:1 ratio and is the center of gravity of the triangle |

| Segment | Concurrent at | Feature |
|---------|---------------|---------|
| Altitudes | Orthocenter | Located inside acute triangles, on a vertex of right triangles, and outside obtuse triangles |

The following phrase can help you remember the points of concurrency:

"**A**ll **o**f **m**y **c**hildren **a**re **b**ringing **i**n **p**eanut **b**utter **c**ookies"

AO—altitudes/orthocenter
MC—medians/centroid
ABI—angle bisectors/incenter
PBC—perpendicular bisectors/circumcenter

## ANGLE AND SEGMENT RELATIONSHIPS IN TRIANGLES

| Angle sum theorem | The sum of the measures of the interior angles of a triangle equals 180°. | $m\angle A + m\angle B + m\angle C = 180°$ |
|---|---|---|
| Exterior angle theorem | The measure of any exterior angle of a triangle equals the sum of the measures of the nonadjacent interior angles. | $m\angle 1 = m\angle 3 + m\angle 4$ |

| Isosceles triangle theorem and its converse | If two sides of a triangle are congruent, then the angles opposite them are congruent.<br><br>If two angles in a triangle are congruent, then the sides opposite them are congruent. | <br>$\angle A \cong \angle B, \overline{AC} \cong \overline{BC}$ |
|---|---|---|
| Equilateral triangle theorem | All interior angles of an equilateral triangle measure 60°. | <br>$\overline{AB} \cong \overline{BC} \cong \overline{AC},$<br>$m\angle A = m\angle B = m\angle C = 60°$ |
| Pythagorean theorem | In a right triangle, the sum of the squares of the legs equals the square of the hypotenuse. | |

# Practice Exercises

1. In $\triangle FGH$, $K$ is the midpoint of $\overline{GH}$. What type of segment is $\overline{FK}$?
   (1) median                (3) angle bisector
   (2) altitude              (4) perpendicular bisector

2. The side lengths of a triangle are 6, 8, and 10. The triangle can be classified as
   (1) equilateral          (3) obtuse
   (2) acutc                (4) right

3. The angle measures of a triangle are $(6x + 6)°$, $(8x - 8)°$, and $(12x)°$. The triangle can be classified as
   (1) obtuse               (3) isosceles
   (2) equilateral          (4) scalene

4. Line $m$ // line $n$, and $\triangle ABC$ is isosceles with $AC = BC$. What is the measure of $\angle 3$?

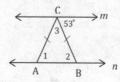

   (1) 53°                  (3) 63.5°
   (2) 60°                  (4) 74°

5. In △*ABC*, $\overline{CY}$ is an angle bisector, m∠*AYC* = 71° and m∠*B* = 23°. What is the measure of angle *A*?

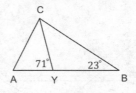

(1) 61°          (3) 76°
(2) 71°          (4) 86°

6. △*RST* is a right triangle with right angle ∠*RST*. △*PSR* is equilateral. Find the measure of angle *T*.

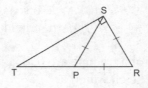

(1) 15°          (3) 30°
(2) 25°          (4) 45°

7. In the figure below, triangle *ABC* is isosceles with *AC* = *BC*, m∠*A* = 70°, and $\overline{AC}$ // $\overrightarrow{BE}$. Find the measure of ∠*CBE*.

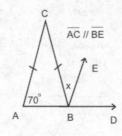

(1) 20°          (3) 35°
(2) 25°          (4) 40°

8. $\overline{JHFK}$ intersects $\overline{IHG}$ at $H$. If m$\angle KFG$ = 156°, m$\angle G$ = 117°, m$\angle JHI$ = $(2x + 15)°$, and m$\angle J$ = $(4x + 6)°$, what is the measure of angle $J$?

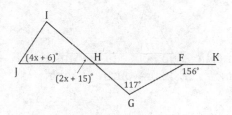

(1) 42°        (3) 62°
(2) 54°        (4) 70°

9. $\angle A$ of $\triangle ABC$ is a right angle, and $\overline{CD}$ is a median. Find the length $CD$ if $AC$ = 9 and $BC$ = 13.

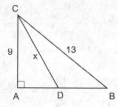

10. In $\triangle ABC$ shown below, m$\angle B$ = 36°, $AB = BC$, and $\overline{AD}$ bisects $\angle BAC$.

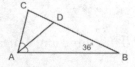

Is $\triangle CAD$ isosceles? Justify your answer.

**11.** $\overline{RTV}$ intersects $\overline{STU}$ at $T$.

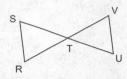

What is the relationship between the sum $(m\angle R + m\angle S)$ and the sum $(m\angle V + m\angle U)$? Explain your reasoning.

**Solutions**

**1.** A median is a segment that joins a vertex of a triangle to the midpoint of the opposite side.

The correct choice is **(1)**.

**2.** The three side lengths satisfy the Pythagorean theorem:

$$a^2 + b^2 = c^2$$
$$6^2 + 8^2 = 10^2$$
$$36 + 64 = 100$$
$$100 = 100$$

Therefore, the triangle is a right triangle.

The correct choice is **(4)**.

**3.**  $6x + 6 + 8x - 8 + 12x = 180$     angle sum theorem
$$26x - 2 = 180$$
$$26x = 182$$
$$x = 7$$

Substitute $x = 7$ to find the three angle measures.

$$(6 \cdot 7 + 6)° = 48°$$
$$(8 \cdot 7 - 8)° = 48°$$
$$(12 \cdot 7)° = 84°$$

Since the triangle has two congruent angles, it must have two congruent sides, so it is isosceles.

The correct choice is **(3)**.

**4.** $m\angle 2 = 53°$ 

alternate interior angles formed by parallel lines are congruent

$m\angle 1 = m\angle 2$

angles opposite congruent sides in a triangle are congruent

$m\angle 1 = 53°$

$m\angle 1 + m\angle 2 + m\angle 3 = 180°$     angle sum theorem
$$53° + 53° + m\angle 3 = 180°$$
$$106° + m\angle 3 = 180°$$
$$m\angle 3 = 74°$$

The correct choice is **(4)**.

**5.** $m\angle AYC = m\angle B + m\angle BCY$ 

exterior angle theorem in $\triangle YBC$

$$71° = 23° + m\angle BCY$$
$$m\angle BCY = 48°$$
$$m\angle ACY = m\angle BCY$$

angle bisector forms 2 congruent angles

$m\angle ACY = 48°$

$m\angle A + m\angle AYC + m\angle ACY = 180°$     angle sum theorem in $\triangle AYC$

$$m\angle A + 71° + 48° = 180°$$
$$m\angle A + 119° = 180°$$
$$m\angle A = 61°$$

The correct choice is **(1)**.

**6.** $m\angle R = m\angle RPS = m\angle PSR = 60°$ 

equilateral triangle theorem

$m\angle RST = 90°$

right angles measure 90°

$m\angle T + m\angle RST + m\angle R = 180°$

angle sum theorem in $\triangle RST$

$$m\angle T + 90° + 60° = 180°$$
$$m\angle T + 150° = 180°$$
$$m\angle T = 30°$$

The correct choice is (**3**).

**7.** The strategy is to first find m$\angle C$, which must be congruent to $\angle CBE$.

$$m\angle ABC = m\angle A = 70° \qquad \text{isosceles triangle}$$
$$\text{theorem}$$

$$m\angle C + m\angle A + m\angle ABC = 180° \qquad \text{triangle angle sum}$$
$$\text{theorem}$$

$$m\angle C + 70° + 70° = 180°$$
$$m\angle C + 140° = 180°$$
$$m\angle C = 40°$$
$$m\angle CBE = m\angle C \qquad \text{congruent alternate}$$
$$\text{interior angles}$$

$$m\angle CBE = 40°$$

The correct choice is (**4**).

**8.** Start by finding m$\angle GHK$ by applying the exterior angle theorem in $\triangle FGH$.

$$m\angle GHF + m\angle FGH = m\angle KFG \qquad \text{exterior angle theorem}$$
$$m\angle GHF + 117° = 156°$$
$$m\angle GHF = 39°$$
$$m\angle JHI = m\angle GHF \qquad \text{vertical angles are}$$
$$\text{congruent}$$

$$(2x + 15)° = 39°$$
$$2x = 24$$
$$x = 12$$
$$m\angle J = (4 \cdot 12 + 6)° \qquad \text{substitute } x = 12$$
$$= 54°$$

The correct choice is (**2**).

**9.** Start by using the Pythagorean theorem in $\triangle ABC$ to find $AB$. From $AB$, calculate $AD$, and then use the Pythagorean theorem again on $\triangle ADC$ to find $CD$.

$$a^2 + b^2 = c^2 \qquad\qquad \text{Pythagorean theorem}$$

$$
\begin{aligned}
AB^2 + AC^2 &= BC^2 \\
AB^2 + 9^2 &= 13^2 \\
AB^2 + 81 &= 169 \\
AB^2 &= 88 \\
AB &= \sqrt{88} \\
&= \sqrt{4 \cdot 22} \\
&= 2\sqrt{22}
\end{aligned}
$$

Median $\overline{CD}$ intersects midpoint $D$, so

$$
\begin{aligned}
AD &= \frac{1}{2} AB \\
AD &= \frac{1}{2} \left( 2\sqrt{22} \right) \\
&= \sqrt{22}
\end{aligned}
$$

Now apply the Pythagorean theorem in $\triangle ADC$:

$$
\begin{aligned}
a^2 + b^2 &= c^2 \\
AD^2 + AC^2 &= CD^2 \\
\left( \sqrt{22} \right)^2 + 9^2 &= CD^2 \\
22 + 81 &= CD^2 \\
103 &= CD^2 \\
CD &= \sqrt{103}
\end{aligned}
$$

**10.** $\triangle BAC$ has $2 \cong$ sides $\overline{AB}$ and $\overline{BC}$, so it is isosceles. The strategy is to find the measures of $\angle C$, $\angle ADC$, and $\angle CAD$.

$$m\angle BAC = m\angle C \quad \text{isosceles triangle theorem in } \triangle ABC$$

$$m\angle BAC + m\angle C + m\angle B = 180° \quad \text{angle sum theorem in } \triangle ABC$$

$$m\angle BAC + \angle BAC + m\angle B = 180° \quad \text{substitute } m\angle BAC \text{ for } m\angle C$$

$$2m\angle BAC + 36° = 180°$$

$$2m\angle BAC = 144°$$

$$m\angle BAC = 72°$$

$$m\angle C = 72°$$

From the angle bisector, we know

$$m\angle CAD = \frac{1}{2} m\angle BAC$$

$$= \frac{1}{2}(72°)$$

$$= 36°$$

Find $m\angle CDA$ by applying the angle sum theorem to $\triangle CAD$.

$$m\angle CDA + m\angle C + m\angle CAD = 180°$$

$$m\angle CDA + 72° + 36° = 180°$$

$$m\angle CDA + 108° = 180°$$

$$m\angle CDA = 72°$$

Both $\angle C$ and $\angle CDA$ measure 72°; therefore, $\overline{AC} \cong \overline{AD}$, and $\triangle CAD$ is isosceles.

**11.** $m\angle RTS + m\angle R + m\angle S = 180°$ from the angle sum theorem. Also, $m\angle VTU + m\angle V + m\angle U = 180°$ for the same reason; therefore,

$$m\angle RTS + m\angle R + m\angle S = m\angle VTU + m\angle V + m\angle U$$

$\angle RTS$ and $\angle VTU$ are congruent vertical angles, so we can subtract their measure from each side of the equation, resulting in

$$m\angle R + m\angle S = m\angle V + m\angle U$$

Therefore, the two sums are equal.

# 3.3 CONSTRUCTIONS

## COPY AN ANGLE

- Given angle $\angle ABC$ and $\overline{DE}$, construct $\angle EDF$ congruent to $\angle ABC$.

  1. Place point on $B$ and make an arc intersecting the angle at $R$ and $S$.
  2. With the same compass opening, place point on $E$ and make an arc intersecting at $T$.
  3. Place point on $R$ and pencil on $S$ and make a small arc.
  4. With the same compass opening, place point on $T$ and make an arc intersecting the previous one at $F$.
  5. $\angle DEF$ is congruent to $\angle ABC$.

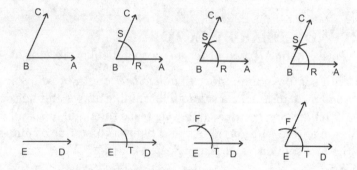

## EQUILATERAL TRIANGLE

- Given segment $\overline{AB}$, construct an equilateral triangle with side length $AB$.

  1. Place point on $A$ and pencil on $B$ and make a quarter circle.
  2. Place point on $B$ and pencil on $A$ and make a quarter circle that intersects the first at $C$.
  3. $\triangle ABC$ is an equilateral triangle.

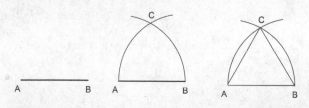

## ANGLE BISECTOR

- Given angle ∠*ABC*, construct angle bisector $\overrightarrow{BD}$.

1. Place point on *B* and make an arc intersecting $\overrightarrow{BA}$ and $\overrightarrow{BC}$ at *R* and *S*.
2. Place point on *R* and make an arc in interior of the angle.
3. With the same compass opening, place point on *S* and make an arc that intersects the previous arc at point *D*.
4. $\overrightarrow{BD}$ is the angle bisector.

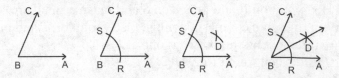

## PERPENDICULAR BISECTOR

- Given segment $\overline{AB}$, construct the perpendicular bisector of $\overline{AB}$.

1. With compass open more than half the length of $\overline{AB}$, place the point at *A* and make a semicircle running above and below $\overline{AB}$.
2. With the same compass opening, place the point at *B* and make a semicircle running above and below $\overline{AB}$ so that it intersects the first semicircle at *R* and *S*.
3. $\overleftrightarrow{RS}$ is the perpendicular bisector of $\overline{AB}$.

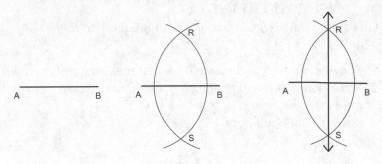

# PERPENDICULAR TO LINE FROM A POINT NOT ON THE LINE

- Given line $\overleftrightarrow{AB}$ and point $P$ not on $\overleftrightarrow{AB}$, construct a line perpendicular to $\overleftrightarrow{AB}$ passing through $P$.

  1. Place point at $P$ and make an arc intersecting $\overleftrightarrow{AB}$ at $R$ and $S$ (extend $\overleftrightarrow{AB}$ if necessary).
  2. Place point at $R$ and make an arc on opposite side of line as $P$.
  3. Place point at $S$ and make an arc intersecting the previous arc at $Q$.
  4. $\overleftrightarrow{PQ}$ is perpendicular to $\overleftrightarrow{AB}$ and passes through $P$.

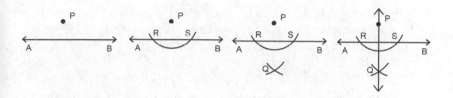

# PERPENDICULAR TO LINE FROM A POINT ON THE LINE

- Given line $\overleftrightarrow{AB}$ and point $P$ on $\overleftrightarrow{AB}$ between $A$ and $B$, construct a line perpendicular to $\overleftrightarrow{AB}$ and passing through $P$.

  1. Place point at $P$ and make an arc intersecting $\overleftrightarrow{AB}$ at $R$ and $S$.
  2. Place point at $R$ and make an arc.
  3. Place point at $S$ and make an arc intersecting the previous arc at $Q$.
  4. $\overleftrightarrow{PQ}$ is perpendicular to $\overleftrightarrow{AB}$ and passes through $P$.

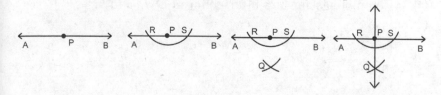

# PARALLEL TO LINE THROUGH POINT OFF THE LINE

- Given $\overleftrightarrow{AB}$ and point $P$ not on $\overleftrightarrow{AB}$, construct a line parallel to $\overleftrightarrow{AB}$ and passing through $P$.

  1. Construct a line passing through $P$ and intersecting $\overleftrightarrow{AB}$ at $R$.
  2. With point at $R$, make an arc intersecting $\overleftrightarrow{PR}$ and $\overleftrightarrow{AB}$ at $S$ and $T$.
  3. With point at $P$ and the same compass opening, make an arc intersecting $\overleftrightarrow{PR}$ at $U$.
  4. With point at $T$, make an arc intersecting at $S$.
  5. With point at $U$ and the same compass opening, make an arc intersecting at $V$.
  6. $\overleftrightarrow{PV}$ is parallel to $\overleftrightarrow{AB}$ and passes through point $P$.

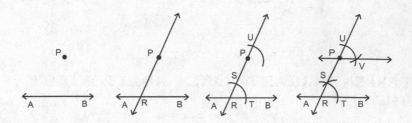

# INSCRIBE A REGULAR HEXAGON OR EQUILATERAL TRIANGLE IN A CIRCLE

  1. Construct circle $P$ of radius $PA$.
  2. Using the same radius as the circle, place point on A and make an arc intersecting circle at $B$.
  3. Place point at B and make an arc intersecting circle at $C$. Continue making arcs intersecting at $D$, $E$, and $F$.

4. Connect each point to form regular hexagon *ABCDEF* or every other point to form equilateral triangle *ACE*.

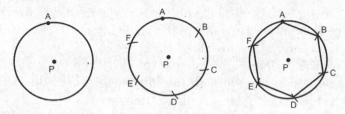

## INSCRIBE A SQUARE IN A CIRCLE

- Given circle *T*

  1. Construct a diameter through the center point *T*.
  2. Construct the perpendicular bisector of the first diameter, which will also be a diameter.
  3. The intersections of the diameters with the circle are the vertices of the square.

  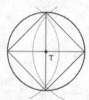

## INSCRIBED CIRCLE

- Given $\triangle ABC$, construct the inscribed circle.

  1. Construct the three angle bisectors of the triangle.
  2. The point of concurrency of the angle bisectors, or incenter $I$, is the center of the inscribed circle.
  3. Construct a line perpendicular to one of the sides of the triangle that passes through the incenter $I$. Label the point of intersection with the side of the triangle $P$.
  4. Construct the inscribed circle with center $I$ and radius $IP$.

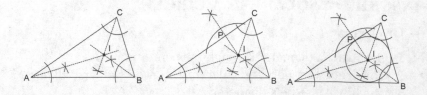

## CIRCUMSCRIBED CIRCLE

- Given $\triangle ABC$, construct the circumscribed circle.

  1. Construct the three perpendicular bisectors of the triangle.
  2. The point of concurrency of the perpendicular bisectors, or circumcenter $P$, is the center of the circumscribed circle.
  3. Construct the circumscribed circle with center $P$ and radius $PA$. (The circle should pass through points $B$ and $C$ as well.)

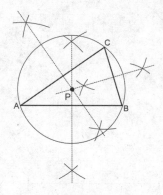

# Practice Exercises

Each of the constructions described in this chapter is a required construction and should be practiced. The constructions in this set of practice problems use the required constructions in various combinations. The first step in each problem is to identify which of the required constructions is called for.

1. Construct a triangle whose angles measure 30°, 60°, and 90°.

2. Given △*JOT*, construct a median from vertex *O*.

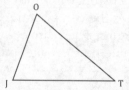

3. Construct the centroid of △*HOT* using a compass and straightedge. Use the compass to confirm your construction by demonstrating the 2:1 ratio. Leave all marks of construction.

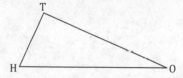

**4.** Given △XYZ, construct the altitude from vertex X.

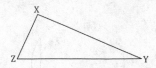

**5.** Given △XYZ, construct a midsegment that intersects sides $\overline{XZ}$ and $\overline{YZ}$.

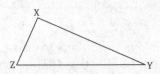

**6.** Given $\overline{AB}$ and $\overline{BC}$, construct parallelogram ABCD.

**7.** Construct a triangle whose angles measure 45°, 45°, and 90° and has $\overline{AB}$ as one of its legs.

A ——— B

**8.** Construct a 15° angle.

9. $\overline{JK}$ and $\overline{LM}$ are chords in circle $P$. Construct the location of point $P$.

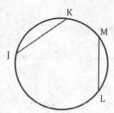

10. Given $\overline{AB}$ and point $P$, construct the dilation of $\overline{AB}$ with a scale factor of 2 and center at $P$.

11. Construct the translation of point $P$ by vector $\overrightarrow{AB}$.

12. $P'$ is the reflection of $P$ over line $m$. Construct line $m$.

P'

P

13. Construct point $Q'$, the rotation of point $Q$ about center $P$ by $\angle ABC$.

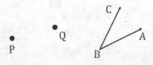

**Solutions**

1. Construct an equilateral triangle, then bisect one of the angles.

2. Construct $A$, the midpoint of $\overline{JT}$, using the "perpendicular bisector" construction. Connect $O$ to $A$. $\overline{OA}$ is a median of $\triangle JOT$.

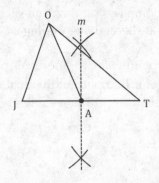

3. The centroid is the point of concurrency of the medians.

- First construct midpoints of two sides of the triangle.
- Then use those midpoints to construct the medians.
- The centroid, $G$, is the point where the medians intersect.

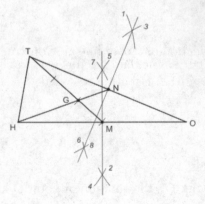

To confirm the centroid divides a median in a 2:1 ratio, measure the distance *GM* and show the distance *TG* is twice *GM*:

- Place the compass point at *M* and make an arc at *G*.
- With the same compass opening, place the point at *G* and make a 2nd arc in the direction of *T*.
- With the same compass opening, you should be able to make a 3rd arc starting at the 2nd and ending at *T*.

**4.** Using the "perpendicular to line from point not on the line" construction, construct a line perpendicular to $\overline{YZ}$ through point *X*. Let *A* be the intersection of the perpendicular line with $\overline{YZ}$. $\overline{AX}$ is the altitude of △*XYZ*.

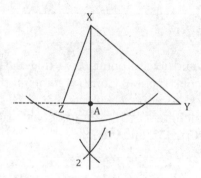

**5.** Using the "perpendicular bisector" construction, construct the midpoints of $\overline{XZ}$ and $\overline{YZ}$. Label these *A* and *B*. $\overline{AB}$ is a midsegment of △*XYZ*.

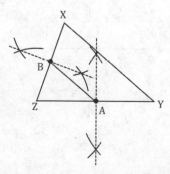

**6.** The steps are

- Extend $\overline{BA}$.
- Use the "parallel line" construction to construct a line parallel to $\overline{BC}$ and through point $A$.
- Measure the length $BC$ by making arc 1 with point at $B$ and pencil at $C$.
- Locate point $D$ by measuring a segment of the same length from point $A$. $D$ is the fourth vertex of parallelogram $ABCD$.

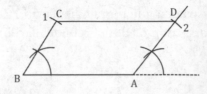

**7.** Extend $\overline{AB}$, and locate point $C$ so that $AB = BC$ using the "copy a segment" construction. Construct line $m$, the perpendicular bisector of $\overline{AC}$. Locate point $D$ so that $AB = BD$ using the "copy a segment" construction. $\triangle ABD$ is an isosceles right triangle with angles measuring $45°$, $45°$, and $90°$.

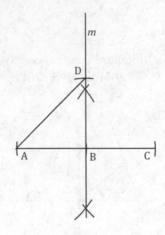

**8.** Construct an equilateral triangle. Bisect one of its angles to get a pair of 30° angles, and then bisect one of those angles to create two 15° angles.

**9.** Construct the perpendicular bisectors of $\overline{JK}$ and $\overline{LM}$. The perpendicular bisector of a chord always passes through the center of the circle, so the intersection of the two perpendicular bisectors is point $P$, the center of the circle.

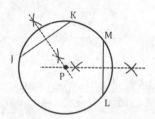

**10.** The distance from a point to the center is multiplied by the scale factor of the dilation, so $PA' = 2PA$ and $PB' = 2PB$. Construct rays $\overrightarrow{PA}$ and $\overrightarrow{PB}$. With the point at $P$, measure the distance $PA$. Move the point to $A$, and measure to same distance to locate $A'$. Repeat for point $B$ and $B'$.

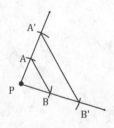

**11.** Use the "parallel to a line through point off the line" construction to construct a ray through $P$ and parallel to $\overleftrightarrow{AB}$. With the point at $A$, measure length $AB$. Move the point to $P$, and measure the same distance along the new ray to locate $P'$. $P'$ is the translation of $P$ by vector $\overrightarrow{AB}$.

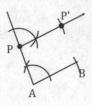

**12.** The line of reflection is the perpendicular bisector of the segment through a point and its image. Construct the perpendicular bisector of $\overleftrightarrow{PP'}$.

**13.** Construct a ray through $P$ and $Q$, and apply the "copy an angle" construction to copy $\angle ABC$ to $\angle PQD$. Then, with the point at $P$, measure the distance $PQ$. Use the same distance to locate point $Q'$ along $\overrightarrow{PD}$. $Q'$ is the rotation of $Q$ about point $P$ by $\angle ABC$.

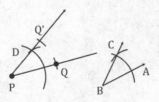

## 3.4 TRANSFORMATIONS

*Transformations* can be thought of as a function that operates on geometric figures. The original figure is called the pre-image and the new figure is called the image. There are four basic transformations to be familiar with—translations, reflections, rotations, and dilations. The first three are rigid motions that do not change the size of the figure and result in a congruent figure. The last one, dilations, can change the size of a figure and results in a similar figure.

## TRANSLATIONS

A *translation* slides a figure from one position to another. Translations can be specified by a vector, with the notation $T_{\overrightarrow{FG}}$. A vector consists of a magnitude and a direction. The magnitude is represented by the length of the vector and the direction is indicated by following the endpoint, point $F$, toward the second point. A translation can also be specified by stating one point and the point it maps to. For example, a translation such that maps $E$ to $G$. The accompanying figure illustrates $T_{\overrightarrow{FG}}(ABCDE) \rightarrow A'B'C'D'E'$. $ABCDE$ is translated a distance $FG$ parallel to the direction of $\overrightarrow{FG}$. The translation is a rigid motion so $ABCDE \cong A'B'C'D'E'$. Also, segments joining corresponding points are congruent and parallel; $\overline{AA'}$ is congruent and parallel to $\overline{BB'}$, $\overline{CC'}$, $\overline{DD'}$, and $\overline{EE'}$.

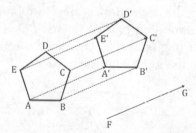

## LINE REFLECTIONS

A *line reflection* flips a figure over a line, so that it appears as a mirror image. The line is called the *line of reflection*. The notation $r_\ell$ indicates a reflection over the line $\ell$. The accompanying figure illustrates the reflection of *ABCD* over line $\ell$. Reflections are a rigid motion, so $ABCD \cong A'B'C'D'$. The line of reflection is also the perpendicular bisector of each segment joining corresponding points in the pre-image and image. Therefore, every point on the pre-image and its corresponding point on the image are equidistant from the line of reflection.

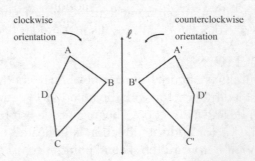

Line reflections do not preserve orientation, which is the direction one must travel to move from one vertex to the next. In the figure, the orientation changes from clockwise to counterclockwise. Noting a change in orientation is one way to identify a reflection.

## POINT REFLECTION

A *point reflection* maps each point in the pre-image to a point such that the center of the reflection is the midpoint of the segment through any corresponding pair of points. $\overline{PQ}$ is the image of $\overline{ST}$ after a reflection through point *R*. Therefore *R* is the midpoint of $\overline{SP}$ and $\overline{TQ}$. A point reflection is equivalent to a 180° rotation.

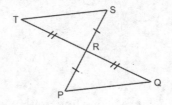

## ROTATIONS

A *rotation* is the spinning of a figure about a pivot point called the *center of rotation*. The angle of rotation is measured counterclockwise unless otherwise specified. The accompanying figure shows *ABCD* rotated about point *O* to *A'B'C'D'*. The notation $R_{C,a}$ is used to specify a rotation, where $a$ is the angle of rotation and $C$ is the center of rotation. Rotations are rigid motions, so $ABCD \cong A'B'C'D'$. The angle formed by any two corresponding points with the center of rotation as the vertex is equal to the angle of rotation. m$\angle AOA'$ = m$\angle BOB'$ = m$\angle COC'$ = m$\angle DOD'$ = angle of rotation.

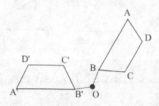

## DILATIONS

A *dilation* is a similarity transformation that enlarges or reduces the size of a figure without changing its shape. Angle measures are preserved, and the lengths of segments in the image are proportional to lengths in the pre-image. The ratio of lengths is called the scale factor. Dilations are specified about a center point. The distance from a point to the center is also multiplied by the scale factor after a dilation. $D_{C,k}$ is a dilation with a center point $C$ and scale factor $k$. The accompanying figure shows the dilation $D_{C,2}(\overline{PQ}) \rightarrow \overline{P'Q'}$.

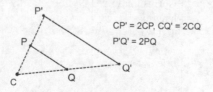

CP' = 2CP, CQ' = 2CQ
P'Q' = 2PQ

# HORIZONTAL AND VERTICAL STRETCH

Two examples of transformations that are neither rigid motions nor similarity transformations are the horizontal stretch and vertical stretch. A *horizontal stretch* will elongate a figure only in the horizontal direction. On the coordinate plane, the x-coordinate of every point will be multiplied by a scale factor. A *vertical stretch* will do the same thing to the y-coordinate.

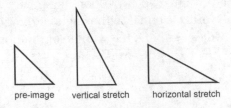

pre-image    vertical stretch    horizontal stretch

## Summary of the Properties of Transformations

| Transformation | What's Preserved | Segments Between Corresponding Points Congruent? | Other Relationships |
|---|---|---|---|
| Reflection $r_{\overline{FE}}(\overline{AB}) \rightarrow \overline{A'B'}$ | Corresponding lengths and angles, parallelism | Not necessarily $AA' \neq BB'$ | $\overline{FE}$ is the $\perp$ bisector of $\overline{BB'}$ and $\overline{AA'}$ $\overline{AA'} \parallel \overline{BB'}$ |
| Rotation $R_{C,100°}(\overline{AB}) \rightarrow \overline{A'B'}$ | Corresponding lengths and angles, parallelism, orientation | Not necessarily $AA' \neq BB'$ | $\overline{AC} \cong \overline{A'C}$ $\overline{BC} \cong \overline{B'C}$ $\angle ACA' \cong \angle BCB'$ |

| Transformation | What's Preserved | Segments Between Corresponding Points Congruent? | Other Relationships |
|---|---|---|---|
| Translation $T_{\overline{UV}}(\overline{AB}) \to \overline{A'B'}$ | Corresponding lengths and angles, slope, parallelism, orientation | Yes $AA' = BB'$ | $\overline{AB} \parallel \overline{A'B'}$ $\overline{AA'} \parallel \overline{BB'}$ |
| Dilation $D_{C,3/2}(\overline{AB}) \to \overline{A'B'}$ | Corresponding angles, slope, and parallelism, orientation | Not necessarily $AA' \neq BB'$ | $\overline{AB} \parallel \overline{A'B'}$ |

## Transformations on the Coordinate Plane

The following set of rules can be used to apply a transformation to a point on the coordinate plane.

| Reflection | Rotation About the Origin |
|---|---|
| $r_{x\text{-axis}}(x, y) \to (x, -y)$ | $R_{90}(x, y) \to (-y, x)$ |
| $r_{y\text{-axis}}(x, y) \to (-x, y)$ | $R_{180}(x, y) \to (-x, -y)$ |
| $r_{y=x}(x, y) \to (y, x)$ | $R_{270}(x, y) \to (y, -x)$ |
| $r_{y=-x}(x, y) = (-y, -x)$ | |

| Translation | Dilation |
|:---:|:---:|
| $T_{h,k}(x, y) \rightarrow (x + h, y + k)$ | $D_{k,\text{origin}}(x, y) \rightarrow (kx, ky)$ |

- Rotations of 180° and 270° can also be found by repeating the rule for a 90° rotation.

## COMPOSITIONS

A **composition of transformation** is a sequence of transformations where the image found after the first transformation becomes the pre-image for the next transformation. The symbol ∘ is used to indicate a composition, as in $T_{3,4} \circ D_{3,\text{origin}}$. The transformation on the right is performed first. The same composition can also be expressed as $T_{3,4}(D_{3,\text{origin}})$.

## SYMMETRY

A figure has *line symmetry* if a line of reflection can be drawn which divides the figure into two congruent mirror images. For example:

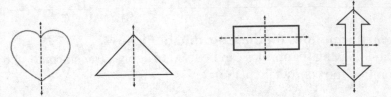

vertical line symmetry          horizontal and vertical
                                line symmetry

A figure has *rotational symmetry* if it can be rotated by some angle $0 < \theta < 360°$ about its center and have every point map to another point on the image. In other words, the image will look identical to the pre-image after the rotation.

The accompanying figure shows rectangle *ABCD* after rotations of 0°, 90°, 180°, 270°, and 360°. Rotations of 0°, 180°, and 360° result in figures that look identical to the original. We say the figure has 180° rotational symmetry—the 0° and 360° rotations are not considered rotational symmetries. Note that the location of individual points has changed. After a 180° rotation, point $A$ is mapped to point $C$, $B$ is mapped to $D$, $C$ is mapped to $A$, and $D$ is mapped to $B$.

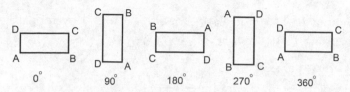

Some figures have more than one rotation that results in an identical figure. An equilateral triangle has 120° and 240° rotational symmetry when rotated about its center.

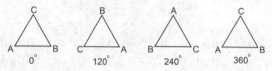

A regular polygon with $n$ sides always has rotational symmetry, with rotations in increments equal to its central angle of $\dfrac{360°}{n}$.

**Example:**

By how many degrees must a regular octagon be rotated so that it maps onto itself? List all the rotations less than 360° that will map an octagon onto itself.

*Solution:*

$n = 8$ for an octagon.

$$\frac{360°}{n} = \frac{360°}{8} = 45°$$

Any rotation between 0° and 360° that is a multiple of 45° will map the octagon onto itself—45°, 90°, 135°, 180°, 225°, 270°, and 315°.

**Example:**

By how many degrees must pentagon $ABCDE$ be rotated about its center to map point $A$ to point $C$?

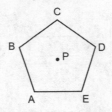

*Solution:*

First calculate the rotation to map one vertex to the adjacent vertex.

$$\frac{360°}{n} = \frac{360°}{5} = 72°$$

Since rotations are counterclockwise, the pentagon must be rotated three increments of 72°, or 216°, to map point $A$ to point $E$.

# Practice Exercises

1. $\triangle ABC$ undergoes a transformation such that its image is $\triangle A'B'C'$. If the side lengths and angles are preserved, but the slopes of the sides are not, the transformation could be a
   (1) reflection or rotation.
   (2) reflection or translation.
   (3) translation or rotation.
   (4) dilation or translation.

2. Which of the following transformations is not a rigid motion?
   (1) reflection
   (2) dilation
   (3) rotation
   (4) translation

3. Which of the following letters has more than one line of symmetry?
   (1) **A**
   (2) **C**
   (3) **H**
   (4) **L**

4. Triangles *AMN* and *APQ* share vertex *A*. A circle can be constructed that has a center at *A* and passes through points *M*, *N*, *P*, and *Q*. If m∠*MAN* = m∠*PAQ*, which of the following is *not* necessarily true?

   (1) ∠*NAQ* is a right angle.
   (2) ∠*M* ≅ ∠*Q*.
   (3) △*APQ* is the image of △*AMN* after a rotation about point *A*.
   (4) △*AMN* and △*AQP* are isosceles.

5. Which transformation is represented in the figure below?

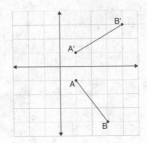

   (1) $R_{origin,\ 90°}$          (3) $r_{x\text{-axis}}$
   (2) $R_{origin,\ 180°}$         (4) $r_{y\text{-axis}}$

6. Which of the following will always map a rectangle onto itself?
   (1) reflection over one of its sides
   (2) reflection over one of its diagonals
   (3) rotation of 180° about its center
   (4) rotation of 90° about one of its vertices

7. The area of a rectangle is 4 cm². The rectangle then undergoes a rotation of 90° about one of its vertices, followed by a dilation with a scale factor of 3, centered at the same vertex as the rotation. What is the area of the image?
   (1) 12 cm²           (3) 32 cm²
   (2) 16 cm²           (4) 36 cm²

8. Regular hexagon ABCDEF is shown in the figure below. Which of the following transformations will map the hexagon onto itself such that point D maps to point E?

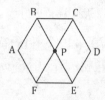

   (1) a 60° rotation about point P followed by a reflection of line $\overline{BE}$
   (2) reflection over line $\overline{BE}$ followed by 60° rotation about point P
   (3) reflection over line $\overline{CF}$ followed by a translation by vector $\overrightarrow{FE}$
   (4) reflection over line $\overline{CF}$ followed by a 120° rotation about point P

9. After a certain transformation the image of △BAD with coordinates B(7, 1) A(0, 3) D(12, 4) is B′(13, 3), A′(6, 5), D′(18, 1). Is the transformation a translation? Explain why or why not.

10. The vertices of △MNO have coordinates M(−5, 2), N(1, 2), P(0, 5). △M′N′P′ is the image of △MNP after a translation. Find the coordinates of N′ and P′ if vertex M is mapped to M′(3, 6).

11. In the figure below, $\triangle DAB \cong \triangle BED \cong \triangle EBC \cong \triangle CFE$. Which of the following sequences of rigid motions will always map $\triangle DAB$ to $\triangle CFE$?

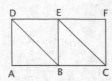

   (1) a translation by vector $\overrightarrow{BE}$ followed by a translation by vector $\overrightarrow{EC}$
   (2) a rotation of 180° about point $B$, followed by a translation by vector $\overrightarrow{CF}$
   (3) a reflection over $\overline{BE}$ followed by a translation by vector $\overrightarrow{BC}$
   (4) a translation by vector $\overrightarrow{AC}$ followed by a reflection over $\overline{FC}$

12. Which of the following will map parallelogram $ABCD$ onto itself?

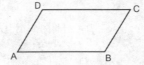

   (1) a rotation of 180° about the point of intersection of the two diagonals of $ABCD$
   (2) a reflection over $\overline{CD}$, followed by a translation along vector $\overrightarrow{DA}$
   (3) a reflection over diagonal $\overline{AC}$
   (4) a reflection over $\overline{BC}$, followed by a translation by vector $\overrightarrow{BA}$

**13.** In the figure below, $\triangle SDR \sim \triangle RCP$. Which of the following compositions of transformations will map $\triangle RCP$ to $\triangle SDR$?

(1) a translation by vector $\overrightarrow{PR}$, followed by a 180° rotation about point $R$, followed by a dilation centered at $R$ with scale factor $\dfrac{RS}{RP}$

(2) a dilation centered at $R$ with scale factor $\dfrac{RS}{RP}$, followed by a reflection over $\overline{RS}$

(3) a translation by vector $\overrightarrow{CR}$, followed by a reflection over $\overline{PC}$, followed by a dilation centered at $R$ with scale factor $\dfrac{RS}{RP}$

(4) a dilation centered at $R$ with scale factor $\dfrac{RP}{DS}$, followed by a translation along vector $\overrightarrow{PD}$, followed by a reflection over $\overline{DR}$

**14.** Graph $\triangle A(-2, 1)$, $B(0, 2)$, $C(-1, 4)$. Graph $\triangle A''B''C''$, the image of $\triangle ABC$ after the transformation $T_{1,3} \circ r_{x=y}$. State the coordinates of $A''B''C''$.

**15.** Parallelogram *STAR* has coordinates $S(-8, -4)$, $T(-5, -2)$, $A(1, -5)$, and $R(-2, -7)$.

    a) Graph and state the coordinates of $S'T'A'R'$, the image of *STAR* after $r_{y=-1}$.

    b) Graph and state the coordinates of $S''T''A''R''$, the image of $S'T'A'R'$ after $T_{7,-2}$.

    c) Name a single reflection that would map $T$ to $T''$.

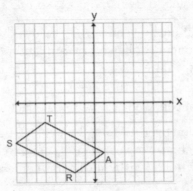

**Solutions**

**1.** A translation always preserves slope, so choices (2), (3), and (4) cannot be correct.

The correct choice is **(1)**.

**2.** Reflections, rotations, and translations are always rigid motions because the image and pre-image are congruent. A dilation may result in an image with a different size, so it is not a rigid motion.

The correct choice is **(2)**.

**3. H** has both horizontal and vertical lines of symmetry.

The correct choice is **(3)**.

**4.** If points $M$, $N$, $P$, and $Q$ all lie on the same circle with center $A$, then $\overline{AM}$, $\overline{AN}$, $\overline{AP}$, and $\overline{AQ}$ are all radii of circle $A$, and $AM = AN = AP = AQ$. The triangles are congruent by SAS and are isosceles, so $\angle M \cong \angle N \cong \angle P \cong \angle Q$ and choices (2) and (4) are true. Since $\angle MAN \cong \angle PAQ$ and $\angle NAP \cong \angle NAP$ by the reflexive property, then $m\angle MAN + m\angle NAP = m\angle NAP + m\angle PAQ$. This is equivalent to $\angle MAP \cong \angle NAQ$. Therefore, $P$ is the rotation of $M$ and $Q$ is the rotation of point $N$ about point $A$, so one triangle is the rotation of the other. Choice (3) is true. The only choice that cannot be proven is choice (1); we do not know the measures of any of the angles and cannot tell if $\angle NAQ$ measures $90°$.

The correct choice is **(1)**.

**5.** Approach this problem by simply applying each transformation and checking if $A(1, -1)$ maps to $A'(1, 1)$ and $B(3, -4)$ maps to $B'(4, 3)$. If the correct image results for both, then the choice is correct.

$R_{origin,\ 90°}(1, -1) \to (1, 1)$      $R_{origin,\ 90°}(3, -4) \to (4, 3)$

correct

$R_{origin,\ 180°}(1, -1) \to (-1, 1)$      $R_{origin,\ 180°}(3, -4) \to (-3, 4)$

$A'$ and $B'$ are incorrect

$R_{x\text{-axis}}(1, -1) \to (1, 1)$      $R_{x\text{-axis}}(3, -4) \to (3, 4)$

$B'$ is incorrect

$r_{y\text{-axis}}(1, -1) \to (-1, -1)$      $r_{y\text{-axis}}(3, -4) \to (-3, -4)$

$A'$ and $B'$ are incorrect

The correct answer is choice **(1)**.

**6.** Since a rectangle has twofold symmetry, a rotation of $180°$ about its center will map it to itself.

The correct choice is **(3)**.

**7.** Under a dilation, the area of a polygon is proportional to the scale factor squared, so the area becomes $3^2$ times larger, or 9 times larger. The new area is $9 \cdot 4$ cm$^2$ = 36 cm$^2$.

The correct choice is **(4)**.

**8.** A reflection over $\overleftrightarrow{BE}$ will map point $D$ to point $F$. Each central angle measures $\frac{360°}{60°}$ = 60°. Therefore, a rotation of 60° will map any vertex to the adjacent vertex in the counterclockwise direction. The 60° rotation will map a point at vertex $F$ to vertex $E$.

The correct choice is **(2)**.

**9.** If the transformation is a translation, then every point in the pre-image would undergo the same translation $(x, y) \rightarrow (x + h, y + k)$. Calculate each $h$ and $k$ by working backwards from the pre-image and image:

$B'(13, 3) = B(7 + h, 1 + k)$    $13 = 7 + h, h = 6$    $3 = 1 + k, k = 2$

$A'(6, 5) = A(0 + h, 3 + k)$    $6 = 0 + h, h = 6$    $5 = 3 + k, k = 2$

$D'(18, 1) = D(12 + h, 4 + k)$    $18 = 12 + h, h = 6$    $1 = 4 + k, k = -3$

The transformation is not a translation because $A$ and $B$ undergo the translation $T_{6,2}$ while $D$ undergoes the translation $T_{6,-3}$.

**10.** First, find the translation rule.

$$T_{h,k} \, M(-5, 2) \rightarrow M'(-5 + h, 2 + k)$$

Using the coordinates $(3, 6)$ for $M'$, solve for $h$ and $k$.

$$-5 + h = 3 \qquad\qquad 2 + k = 6$$
$$h = 8 \qquad\qquad\qquad k = 4$$

The translation is $T_{8,4}$. Now apply that translation to $N$ and $P$.

$$T_{8,4} \, N(1, 2) \rightarrow N'(1 + 8, 2 + 4) \rightarrow N'(9, 6)$$
$$T_{8,4} \, P(0, 5) \rightarrow P'(0 + 8, 5 + 4) \rightarrow P'(8, 9)$$

**11.** The figure shows the $\triangle DAB$ after a 180° rotation about $B$. A translation by $\overrightarrow{CF}$ will then slide the image up and map it to $\triangle CFE$.

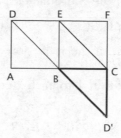

The correct choice is (**2**).

**12.** Rotating 180° about the point of intersection of the diagonals will map: $A \rightarrow C$, $B \rightarrow D$, $C \rightarrow A$, and $D \rightarrow B$.

The other choices will not map $ABCD$ onto itself. Choices (2), (3), and (4) result in the following:

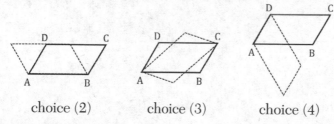

choice (2)      choice (3)      choice (4)

The correct choice is (**1**).

**13.** The figures show the figure after the translation by $\overrightarrow{PR}$, and then the rotation of 180° about $R$. The final dilation will map $C''$ to $D$ and $R''$ to $S$.

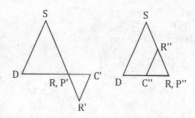

The correct choice is (**1**).

**14.** In a composition always do the rightmost transformation first. For $T_{1,3} \circ r_{x=y}$, start with $r_{y=x}$. For a reflection over the line $y = x$, switch the order of the coordinates.

$r_{y=x}: A(-2, 1) \rightarrow A'(1, -2)$   $B(0, 2) \rightarrow B'(2, 0)$   $C(-1, 4) \rightarrow C'(4, -1)$

Now use the image from the reflection as the pre-image of the translation.

For a translation $T_{1,3}$, add 1 to each $x$-coordinate and 3 to each $y$-coordinate.

$T_{1,3}: A'(1, -2) \rightarrow A''(2, 1)$   $B'(2, 0) \rightarrow B''(3, 3)$   $C'(4, -1) \rightarrow C''(5, 2)$

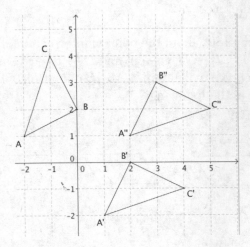

**15.** a) The $x$-coordinates of each point in *STAR* are unchanged after a reflection over the line $y = -1$. To find the $y$-coordinates, count the vertical distance from the point to $y = -1$, then continue by the same distance on the other side of $y = -1$. $S(-8, -4)$ is 3 units below $y = -1$.

The $y$-coordinate of $S'$ is $-1 + 3 = 2$.        $S'(-8, 2)$

$T(-5, -2)$ is 1 unit below $y = -1$.

The $y$-coordinate of $T'$ is $-1 + 1 = 0$. $\qquad$ $T'(-5, 0)$

$A(1, -5)$ is 4 units below $y = -1$.

The $y$-coordinate of $A'$ is $-1 + 4 = 3$. $\qquad$ $A'(1, 3)$

$R(-2, -7)$ is 6 units below $y = -1$.

The $y$-coordinate of $R'$ is $-1 + 6 = 5$. $\qquad$ $R'(-2, 5)$

$S'(-8, 2)$ $T'(-5, 0)$ $A'(1, 3)$ $R'(-2, 5)$

b) Apply the translation $T_{7,-2}$ by adding 7 to each $x$-coordinate and subtracting 2 from each $y$-coordinate.

$$S''(-1, 0)\ T''(2, -2)\ A''(8, 1)\ R''(5, 3)$$

c) To reflect $T(-5, -2)$ to $T'''(2, -2)$, note that the $y$-coordinates are the same, so we need a reflection over a vertical line that is halfway between $x = -5$ and $x = 2$.

$$\frac{1}{2}(-5 + 2) = -1\frac{1}{2}$$

A reflection over the line $x = -1\frac{1}{2}$ will map $T(-5, -2)$ to $T'''(2, -2)$.

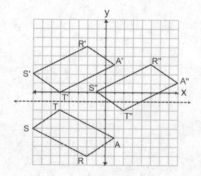

## 3.5 TRIANGLE CONGRUENCE

### CONGRUENCE

Two figures are congruent if one figure can be mapped onto another by a sequence of rigid motions. The congruent figures will have exactly the same size and shape, which means for polygons that all *corresponding* pairs of sides and angles are congruent.

A congruence statement will match up the corresponding pairs of congruent parts.

For example, from the statement $\triangle BUS \cong \triangle CAR$ we can conclude:

$$\angle B \cong \angle C \qquad \overline{BU} \cong \overline{CA}$$
$$\angle U \cong \angle A \qquad \overline{US} \cong \overline{AR}$$
$$\angle S \cong \angle R \qquad \overline{SB} \cong \overline{RC}$$

Once two figures have been proven congruent, we know all pairs of parts are congruent. In triangles, this theorem is abbreviated by CPCTC—corresponding parts of congruent triangles are congruent.

### PROVING POLYGONS CONGRUENT BY TRANSFORMATIONS

Polygons can be proven congruent by identifying a sequence of rigid motions that map one onto the other. One method is to show each vertex of the pre-image maps to a corresponding vertex of the image using the same transformation, or sequence of transformations. The angle and segment relationships between pre-image, image, lines of reflection, center of rotation, and translation vectors can help establish specific transformations that map one point onto another:

- If the segments joining corresponding vertices of the two triangles are congruent and parallel, then one triangle is a translation of the other.
- If a single line is the perpendicular bisector of segments formed by corresponding vertices, then one triangle is the reflection of the other.

- If angles formed by corresponding vertices and a center point are all congruent, and corresponding distances to the center point are equal, then one triangle is a rotation of the other.

For example, given $\overline{ADBE}$, $\overline{AD} \cong \overline{BE}$, $\overline{AD} \cong \overline{CF}$, $\overline{AD} \,//\, \overline{CF}$, we can show that $\triangle ABC \cong \triangle DEF$ because one is a translation of the other. Point $A$ translates to point $D$ along segment $\overline{ADBE}$. $B$ translates to $E$ by the same distance along the same segment, and $C$ translates by the same distance along a parallel segment. Each point in $\triangle ABC$ is mapped to a corresponding point in $\triangle DEF$ by the same translation (distance $AD$ along a vector parallel to $\overline{ADBE}$). Therefore, $\triangle DEF$ is a translation of $\triangle ABC$, and the two triangles are congruent because translations are a rigid motion.

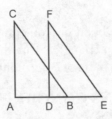

## TRIANGLE CONGRUENCE POSTULATES

We know two polygons are congruent if all pairs of corresponding sides and angles are congruent. When proving two triangles congruent we do not need to prove all three pairs of sides and all three pairs of angles congruent. There are five shortcuts commonly used—the SAS, SSS, ASA, AAS, and the HL criteria. Each criteria requires only a total of three pairs of parts to be congruent for us to conclude the triangles are congruent.

| Criteria | What It Looks Like |
|---|---|
| **SAS**—two pairs of sides and the included angles are congruent. | |
| **SSS**—three pairs of sides are congruent. | |
| **ASA**—two pairs of angles and the included sides are congruent. | |
| **AAS**—two pairs of angles and the nonincluded sides are congruent. | |
| **HL**—the hypotenuses and one pair of legs are congruent in a pair of right triangles. | |

Once a pair of triangles has been proven congruent using one of these postulates, then CPCTC can be used to justify why any of the remaining pairs of corresponding parts are congruent.

You do not need to limit yourself strictly to a rigid motion or a congruence postulate. Sometimes the given information will lead to a combination of these approaches.

## STRATEGIES FOR WRITING CONGRUENCE PROOFS

1. Mark congruent parts on the figure using given information and anything else you can conclude from the figure.
2. Look for the following common relationships:
   a. Vertical angles
   b. Shared sides and angles
   c. Linear pairs
   d. Supplementary and complementary angles
   e. Parallel lines and perpendicular lines
   f. Midpoints and bisectors of segments
   g. Angle bisectors
   h. Altitudes, medians, and perpendicular bisectors in triangles
   i. Isosceles triangle theorem, exterior angle theorem, and triangle angle sum theorem
3. Statements should always refer to specific named parts of the figure (such as $\triangle ABC$, $\overline{AB}$, $\angle C$). Reasons may only involve definitions, postulates, and theorems. Never name a specific part of the figure in the reasons column.
4. There must be a line in your proof for each part of the postulate you use. You can label each line that corresponds to a part of the postulate with (A) or (S), or use checkboxes:

| S | A | S |
|---|---|---|
| ✔ | ✔ | ✔ |

## OVERLAPPING TRIANGLES

Angle addition or segment addition can be used to show parts are congruent when triangles overlap. For example, if given $\overline{ABCD}$ and $\overline{AB} \cong \overline{CD}$, we can prove $\overline{AC} \cong \overline{BD}$ by stating that $\overline{BC}$ is congruent to itself by the reflexive property, then adding the two congruence statements:

| | |
|---|---|
| $\overline{BC} \cong \overline{BC}$ | reflexive property |
| $\overline{AC} \cong \overline{BD}$ | given |
| $\overline{AB} + \overline{BC} \cong \overline{BC} + \overline{CD}$ | addition property |
| $\overline{AC} \cong \overline{BD}$ | partition property |

Another strategy is to sketch the two triangles separately. By pulling them apart, you may more easily see the parts required to complete the proof.

# Practice Exercises

1. $A'B'C'D'$ is the image of $ABCD$ after a rotation about the origin, and $A''B''C''D''$ is the image of $A'B'C'D'$ after a translation. Which of the following must be true?

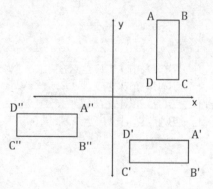

(1) All three figures are rectangles.
(2) $ABCD$ is congruent to $A'B'C'D'$, but not congruent to $A''B''C''D''$.
(3) $ABCD$ is congruent to $A''B''C''D''$, but not congruent to $A'B'C'D'$.
(4) $ABCD$ is congruent to both $A'B'C'D'$ and $A''B''C''D''$.

**2.** Which triangle congruence criteria can be used to prove the two triangles are congruent?

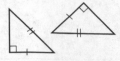

(1) SAS          (3) SSS

(2) ASA        (4) HL

**3.** Which of the following pieces of information would *not* allow you to conclude that $\triangle ACD \cong \triangle ACB$?

(1) $\overline{AC}$ bisects $\angle BCD$ and $\overline{CD} \cong \overline{CB}$.

(2) $\overline{AC}$ bisects $\angle BCD$ and $\overline{AD} \cong \overline{AB}$.

(3) $\overline{AC}$ bisects $\angle BCD$ and $\angle BAD$.

(4) $\overline{AC}$ bisects $\angle BAD$ and $\overline{AD} \cong \overline{AB}$.

**4.** Which of the following is sufficient to prove $\triangle SMA \cong \triangle S'M'A'$?

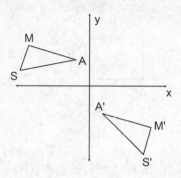

(1) There is a sequence of rigid motions that maps
   $\angle M$ to $\angle M'$ and $\angle S$ to $\angle S'$.
(2) There is a sequence of rigid motions that maps
   $\angle M$ to $\angle M'$ and $\overline{MS}$ to $\overline{M'S'}$.
(3) There is a sequence of rigid motions that maps
   $\angle S$ to $\angle S'$ and $\overline{MS}$ to $\overline{M'S'}$.
(4) There is a sequence of rigid motions that maps
   $\angle A$ to $\angle A'$ and point $A$ to point $A'$.

**5.** Given: $\overline{AB} \parallel \overline{CD}$, $\overline{AB} \cong \overline{CD}$
   Prove: $\triangle ABE \cong \triangle CDE$

**6.** Given: $\overline{BC}$ bisects $\overline{AD}$ at $E$, $\angle A \cong \angle D$
   Prove: $\triangle ABE \cong \triangle DCE$

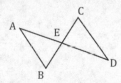

7.  *ABCDEF* is a regular hexagon. State a sequence of rigid motions that could be used to justify why △*ABC* is congruent to △*AFE*.

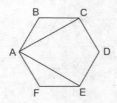

8.  In the accompanying figure of △*FGH* and △*GIJ*, there is a sequence of rigid motions that maps *F* to *G*, *G* to *I*, and *H* to *J*. Explain why ∠*H* must be congruent to ∠*J*.

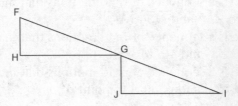

9.  Given $\overline{RPB}$ and line *m* is the perpendicular bisector of $\overline{TJ}$ and $\overline{RB}$, explain why △*TRP* ≅ △*JBP* and why $\overline{TR}$ ≅ $\overline{JB}$.

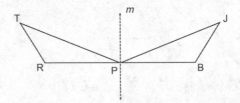

**10.** Given: △WXY and perpendicular bisector $\overline{YZ}$
Explain why △WZY ≅ △XZY in terms of a rigid motion.

**11.** Given: $\overline{AB} \cong \overline{AD}$, $\overline{AC} \cong \overline{AE}$, m∠BAD = 60°, and m∠CAE = 60°
Prove: △BAC ≅ △DAE in terms of a rigid motion

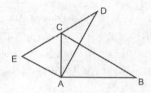

**12.** Given: line $m$ is the perpendicular bisector of $\overline{FX}$, $\overline{GY}$, and $\overline{HZ}$
Prove: △XYZ ≅ △FHG

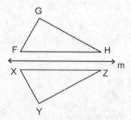

## Solutions

**1.** Both rotations and translations are rigid motions, and any sequence of rigid motions will result in a congruent image. Therefore, $ABCD$ is congruent to both $A'B'C'D'$ and $A''B''C''D''$.

The correct answer choice is (**4**).

**2.** Both triangles are right triangles, a corresponding pair of legs are congruent, and the two hypotenuses are congruent. Therefore HL applies. SAS is not correct here because the angle does not lie between the two congruent sides.

The correct choice is (**4**).

**3.** The shared side $\overline{AC}$ is a pair of congruent sides. Check each choice to find one that does *not* provide enough information to complete one of the congruent postulates.

Choice 1: The angle bisector gives $\angle DCA \cong \angle BCA$, and SAS applies.

Choice 2: The angle bisector gives $\angle DCA \cong \angle BCA$, and we have two sides and an angle, but the angle is not the included angle. None of the congruence postulates apply.

Choice 3: The angle bisector gives $\angle DCA \cong \angle BCA$ and $\angle DAC \cong \angle BAC$. ASA applies.

Choice 4: The angle bisector gives $\angle DAC \cong \angle BAC$, and SAS applies.

The correct choice is (**2**).

**4.** Choice 1 maps two consecutive angles onto corresponding angles, as well as $M$ to $M'$ and $S$ to $S'$. This means $\overline{SM} \cong \overline{S'M'}$ because a sequence of rigid motions maps the endpoints of $\overline{SM}$ onto corresponding endpoints. $\overline{SM}$ is the included side between the congruent angles, therefore, $\triangle SMA \cong \triangle S'M'A'$ by the SAS postulate.

The correct choice is (**1**).

**5.**

| Statement | Reason |
|---|---|
| 1. $\overline{AB} \cong \overline{CD}$, $\overline{AB} \parallel \overline{CD}$ | 1. Given |
| 2. $\angle A \cong \angle C$ $\angle B \cong \angle D$ | 2. Alternate interior angles formed by the parallel lines are congruent |
| 3. $\triangle ABE \cong \triangle CDE$ | 3. ASA |

**6.**

| Statement | Reason |
|---|---|
| 1. $\overline{BC}$ bisects $\overline{AD}$ at $E$ | 1. Given |
| 2. $E$ is the midpoint of $\overline{AD}$ | 2. A bisector intersects a segment at its midpoint |
| 3. $\overline{AE} \cong \overline{ED}$ | 3. A midpoint divides a segment into two congruent segments |
| 4. $\angle AEB \cong \angle DEC$ | 4. Vertical angles are congruent |
| 5. $\angle A \cong \angle D$ | 5. Given |
| 6. $\triangle ABE \cong \triangle DCE$ | 6. ASA |

**7.** $\triangle ABC$ is mapped to $\triangle AFE$ by a rotation about point $A$ followed by a reflection over $\overline{AE}$. The angle of rotation is $\angle CAE$. From the figure each of the diagonals $\overline{AC}$, $\overline{AE}$, and $\overline{CE}$ are congruent, so $\triangle ACE$ is equilateral, and its angles measure 60°. The angle of rotation is 60°. The sequence of rigid motions is a 60° rotation about point $A$ followed by a reflection over $\overline{AE}$.

**8.** The two triangles are congruent because there exists a rigid motion that maps each vertex of $\triangle JGH$ to a corresponding vertex in $\triangle GIJ$. Since the triangles are congruent, all pairs of corresponding parts are congruent (CPCTC). $\angle G$ and $\angle H$ are congruent corresponding angles.

**9.** Since line $m$ is the perpendicular bisector of both $\overline{TJ}$ and $\overline{RB}$, it is a line of reflection that maps $T$ to $J$ and $R$ to $B$. The same reflection will map $P$ to itself because it lies on the line of reflection. A single reflection maps $\triangle TRP$ onto $\triangle JBP$, and reflections are rigid motions, so the two triangles must be congruent. $\overline{TR} \cong \overline{JB}$ because corresponding parts of congruent triangles are congruent.

**10.** Point $Y$ is the image of itself after a reflection over $\overline{YZ}$ because $Y$ lies on the line of reflection. Point $Z$ is the image of itself for the same reason. $X$ is the image of point $W$ after a reflection over $\overline{YZ}$ because a line of reflection is the perpendicular bisector of the segment joining the pre-image and the image after the reflection. Therefore, the triangles are congruent because one is mapped to the other by a reflection over $\overline{YZ}$, which is a rigid motion.

**11.** $\overline{AB} \cong \overline{AD}$; therefore, $D$ is the image of $B$ after a 60° rotation about $A$. $\overline{AC} \cong \overline{AE}$; therefore, $C$ is the image of $E$ after a 60° rotation about $A$. Point $A$ will map to itself after the same rotation because it is the center of rotation. Therefore, $\triangle BAC$ is the image of $\triangle DAE$ after a 60° rotation about $A$. $\triangle BAC \cong \triangle DAE$ because rotations are rigid motions.

**12.** A perpendicular bisector of a segment is also the line of reflection that maps one endpoint to the other. Therefore, after a reflection over line $m$, $X$ is the image of $F$, $Z$ is the image of $H$, and $Y$ is the image of $G$. $\triangle FHG$ is mapped to $\triangle XZY$ by a reflection, which is a rigid motion; therefore, $\triangle XZY \cong \triangle FHG$.

## 3.6 COORDINATE GEOMETRY

### COORDINATE GEOMETRY FORMULAS

Given two points $(x_1, y_1)$ and $(x_2, y_2)$:

- The distance between the points is $\sqrt{(x_2 - x_1)^2 + (y_2 - y_1)^2}$

- The midpoint of the segment joining the points is
$$\left( \frac{x_1 + x_2}{2}, \frac{y_1 + y_2}{2} \right)$$

- The slope of the segment joining the points is $\dfrac{y_2 - y_1}{x_2 - x_1}$, or $\dfrac{rise}{run}$

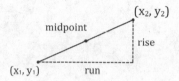

### DIVIDING A SEGMENT PROPORTIONALLY

A directed segment is one that has a specified starting point and ending point. We can divide a directed segment into parts that are in any given ratio using the two proportions ratio $= \dfrac{x - x_1}{x_2 - x}$ and ratio $= \dfrac{y - y_1}{y_2 - y}$.

For example, given $J(1, -2)$ and $K(11, 3)$, find the coordinate $(x, y)$ of point $L$ that divides $\overline{JK}$ in a 2:3 ratio.

J(1, -2)    L(x,y)    K(11, 3)

- Use the coordinates of $J$ for point 1 and the coordinates of $K$ for point 2.

$$x_1 = 1 \qquad x_2 = 11 \qquad y_1 = -2 \text{ and } y_2 = 3$$

- Apply the formula for the $x$-coordinate.

ratio $= \dfrac{x - x_1}{x_2 - x}$

$$\frac{2}{3} = \frac{x-1}{11-x}$$
$$2(11 - x) = 3(x - 1)$$
$$22 - 2x = 3x - 3$$
$$25 = 5x$$
$$x = 5$$

- Repeat the process for the $y$-coordinate.

ratio $= \dfrac{y - y_1}{y_2 - y}$

$$\frac{2}{3} = \frac{y-(-2)}{3-y}$$
$$3y + 6 = 6 - 2y$$
$$5y = 0$$
$$y = 0$$

Point $L(5, 0)$ divides $\overline{JK}$ in a 2:3 ratio.

## AREA AND PERIMETER

The lengths found with the distance formula can be used to calculate the perimeter and area of figures. If the figure is irregular, three strategies can be used:

- Divide the figure into shapes whose areas can be calculated easily (squares, rectangles, triangles, trapezoids, and circles).
- Sketch a bounding rectangle around the figure. Calculate the area of the rectangle, and then subtract the area of all the triangles that fall outside the figure but with the rectangle.
- If the figure has curves, estimate the area by modeling the curved sides with straight segments. The more segments that are used to model a curve, the more accurate the result will be.

**Example:**

Find the area of polygon $ABCDE$, with vertices $A(-3, -5)$, $B(2, -5)$, $C(5, 3)$, $D(-2, 5)$, and $E(-5, -1)$.

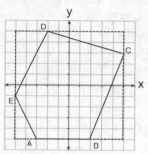

*Solution:*

Sketch the bounding rectangle in around $ABCDE$.

The length is 10 and the width is 10, giving an area of 100.

The triangles have areas:

$$\text{upper left triangle} \quad = \frac{1}{2}(3 \cdot 6) = 9$$

$$\text{upper right triangle} = \frac{1}{2}(7 \cdot 2) = 7$$

$$\text{lower left triangle} \quad = \frac{1}{2}(4 \cdot 2) = 4$$

$$\text{lower right triangle} = \frac{1}{2}(3 \cdot 8) = 12$$

The area of $ABCDE = 100 - 9 - 7 - 4 - 12 = 68$

## COLLINEARITY

Three points are **collinear** if the slopes between any two pairs are equal. For example, points $A$, $B$, and $C$ are collinear if the slope of $\overline{AB}$ equals the slope of $\overline{BC}$.

## EQUATIONS OF LINES

- The slopes of parallel lines are equal.
- The slopes of perpendicular lines are negative reciprocals. (If the slope of line $m$ is $\frac{2}{3}$, then the slope of any line perpendicular to $m$ is $-\frac{3}{2}$.)

- The equation of a line in slope-intercept form is $y = mx + b$ where $m$ is the slope and $b$ is the $y$-intercept. To graph the line, plot a point on the $y$-axis at the $y$-intercept. From that point, plot additional points using the rise and run from the slope.
- The equation of a line in point-slope form is $y - y_1 = m(x - x_1)$ where $m$ is the slope and $(x_1, y_1)$ are the coordinates of any point on the line. To graph the line, plot the first point at $(x_1, y_1)$. From that point, plot additional points using the rise and run from the slope.

Strategy for writing the equation of a line in slope-intercept form:

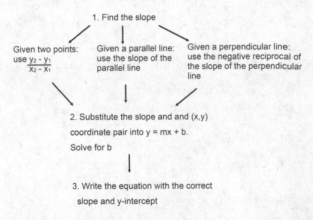

## TRANSFORMATIONS AND LINES

### Translations and Dilations

Translations and dilations preserve slope, so the slope of the image will be the same as the slope of the pre-image.

To translate or dilate a line given its equation,

1. Choose any point on the line (the $y$-intercept is often an easy choice).
2. Apply the translation or dilation to that point.
3. Find the equation of the line that has the same slope as the original line and passes through the transformed point.

*Rotations*

Rotations of 90° will result in a line perpendicular to the original, so the slope will be the negative reciprocal. To write the equation of a line after a 90° rotation, use the same procedure for translations and dilations, except use the negative reciprocal of the slope.

## EQUATION OF THE CIRCLE

### Center Radius Form of the Equation of a Circle

$(x - h)^2 + (y - k)^2 = r^2$ where the center has coordinates $(h, k)$ and radius has length $r$.

- To graph a circle, first identify the center and radius from the equation. Plot a point at the center. Then plot points up, down, left, and right a distance $r$ from the center.

**Example:**

Graph the equation $(x - 2)^2 + (y + 1)^2 = 9$.

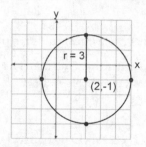

The center is located at $(2, -1)$, and $r^2 = 9$, so $r = 3$. We plot the center point at $(2, -1)$; then plot points up, down, right, and left 3 units from the center. Use these four points as a guide to complete the circle.

### General Form of the Equation of a Circle

$$x^2 + y^2 + Cx + Dy + E = 0$$

To find the coordinates of the center and the radius from the general form of the equation, you will need to convert it to the center–radius form using the following procedure:

1. Group the $x$-terms and $y$-terms on one side of the equation, and the constant on the other side of the equation.
2. Complete the square with the $x$-terms, and then complete the square with the $y$-terms.

**Example:**
- Find the coordinates of the center and the length of the radius of a circle whose equation is $x^2 + 4x + y^2 - 6y + 7 = 0$.

*Solution:*
Bring the constant term to the right.

$$x^2 + 4x + y^2 - 6y = -7$$

The coefficient of $x$ is 4, so a constant term of $\left(\dfrac{4}{2}\right)^2$, or 4, is needed to complete the square with the $x$-terms. The coefficient of $y$ is $-6$, so a constant term of $\left(\dfrac{-6}{2}\right)^2$, or 9, is needed to complete the square with the $y$-terms.

$$x^2 + 4x + 4 + y^2 - 6y + 9 = -7 + 4 + 9$$
$$(x + 2)^2 + (y - 3)^2 = 6$$

The center has coordinates $(-2, 3)$ and the radius has a length of $\sqrt{6}$.

# Practice Exercises

1. Points $A(2, -1)$ and $B(8, -3)$ lie on line $m$. After a rotation of $90°$ about the origin, the images of $A$ and $B$ are $A'$ and $B'$. If $A'$ and $B'$ lie on line $n$, what is the equation of line $n$?

   (1) $y = -3x + 2$  (3) $y = 3x - 1$

   (2) $y = -\dfrac{1}{3}x - \dfrac{1}{3}$  (4) $y = \dfrac{1}{3}x + 6$

2. What is the equation of the line $6x + 2y = 12$ after a dilation by a scale factor of 5?

   (1) $y = -3x + 30$  (3) $y = -15x + 30$

   (2) $y = -3x + 6$  (4) $y = -15x + 6$

3. Which of the following lines is perpendicular to the line $x + 4y = 8$?

   (1) $y = -\dfrac{1}{4}x + 2$  (3) $y = \dfrac{1}{4}x + 2$

   (2) $y = 4x + 3$  (4) $y = -4x + 3$

4. Which of the following is the equation of a line parallel to $2x + 3y + 6 = 0$ and passes through the point $(6, 1)$?

   (1) $y = -\dfrac{2}{3}x + 5$  (3) $y = -2x + 13$

   (2) $y = 2x - 11$  (4) $y = -\dfrac{2}{3}x + 3$

5. What are the coordinates of the midpoint of a segment whose endpoints have coordinates (3, 1) and (15, –7)?
   (1) (27, –15)          (3) (6, –4)
   (2) (–6, 4)            (4) (9, –3)

6. The diameter of a circle has endpoints with coordinates (4, –1) and (8, 3). Which of the following is an equation of the circle?

   (1) $(x - 2)^2 + (y - 1)^2 = 8$      (3) $(x - 6)^2 + (y - 1)^2 = 8$
   (2) $(x - 2)^2 + (y - 1)^2 = 32$     (4) $(x - 6)^2 + (y - 1)^2 = 32$

7. Are the segments $\overline{AB}$ and $\overline{TU}$ congruent, given coordinates $A(1, 4)$, $B(-3, 6)$, $T(2, 5)$, and $U(4, 1)$? Justify your answer.

8. Find the coordinates of the point $W$ that divides directed segment $\overline{UV}$ in a 1:5 ratio, given coordinates $U(-3, 7)$ and $V(9, 1)$.

9. Point $A$ has coordinates $(-2, 7)$ and point $B$ has coordinates $(6, 3)$. Line $m$ has the property that every point on the line is equidistant from points $A$ and $B$. Find the equation of line $m$.

10. A circle is described by the equation $x^2 + 6x + y^2 - 12y + 25 = 0$. Find the radius of the circle and the coordinates of its center.

11. Circle $P$ has a center $P(4, -5)$ and a radius with length $\sqrt{65}$. Does the point $A(8, 2)$ lie on circle $P$? Justify your answer.

12. Parallelogram $ABCD$ has coordinates $A(2, -1)$, $B(5, 1)$, $C(a, b)$, and $D(3, 4)$. Write the equation of the line that contains side $\overline{CD}$.

**13.** Estimate the area of the ellipse shown in the accompanying figure by modeling it as the sum of 5 rectangles.

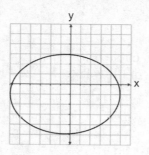

**14.** Ben wants to estimate the area of the curved region shown. He would like to do so by calculating the average of the areas of two different circles.

a) Graph the two circles that Ben could use.

b) Using the two circles from part (a), estimate the area of the curved region. Each grid unit represents 1 centimeter. Round to the nearest 1 cm².

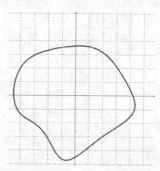

**15.** $\triangle SKY$ has coordinates $S(0, 2)$, $K(6, 0)$, $Y(4, 4)$.

a) Find the equation of the perpendicular bisector to side $\overline{SK}$.

b) Show that the perpendicular bisector is also an altitude.

**Solutions**

**1.** Applying a 90° rotation, $A$ maps to $A'(1, 2)$ and $B$ maps to $B'(3, 8)$.

$$\text{Slope } \overrightarrow{A'B'} = \frac{y_2 - y_1}{x_2 - x_1}$$

$$= \frac{8 - 2}{3 - 1}$$

$$= 3$$

Write the equation of the line using the point-slope form and the coordinates of $A'$ for $x_1$ and $y_1$.

$$y - y_1 = m(x - x_1)$$
$$y - 2 = 3(x - 1)$$
$$y - 2 = 3x - 3$$
$$y = 3x - 1$$

The correct choice is **(3)**.

**2.** A dilation with a positive scale factor will preserve slope, but the distance of each point from the center will be multiplied by the scale factor. The strategy is to rewrite the equation in $y = mx + b$ form. The dilated line will have the same slope, but the $y$-intercept will be multiplied by the scale factor.

$$6x + 2y = 12$$
$$2y = -6x + 12$$
$$y = -3x + 6$$

The $y$-intercept of the original line is 6, and the $y$-intercept of the dilated line is $6 \cdot 5$, or 30.

$$y = -3x + 30$$

The correct choice is **(1)**.

**3.** Perpendicular lines have negative reciprocal slopes. Rewrite the equation in slope-intercept form to identify the slope.

$$x + 4y = 8$$
$$4y = -x + 8$$
$$y = -\frac{1}{4}x + 2$$

The negative reciprocal of $-\frac{1}{4}$ is 4, so the perpendicular line must have a slope of 4. The $y$-intercept can be any value since a particular perpendicular line was not specified. The only line with a slope of 4 is $y = 4x + 3$.

The correct choice is **(2)**.

**4.** Parallel lines have equal slopes, so the first step is to rewrite the equation in slope-intercept form to help identify the slope.

$$2x + 3y + 6 = 0$$
$$3y = -2x - 6$$
$$y = -\frac{2}{3}x - 2$$

The slope is $-\frac{2}{3}$.

Now substitute $x = 6$, $y = 1$, and the slope into $y = mx + b$ and solve for the new $y$-intercept.

$$y = mx + b$$
$$y = -\frac{2}{3}x + b$$
$$1 = -\frac{2}{3}(6) + b$$
$$1 = -4 + b$$
$$b = 5$$

The equation is $y = -\frac{2}{3}x + 5$.

The correct choice is **(1)**.

**5.** Apply the midpoint formula:

$$x_{MP} = \frac{1}{2}(x_1 + x_2) \qquad y_{MP} = \frac{1}{2}(y_1 + y_2)$$

$$x_{MP} = \frac{1}{2}(3 + 15) \qquad y_{MP} = \frac{1}{2}(1 + (-7))$$

$$x_{MP} = \frac{1}{2}(18) \qquad y_{MP} = \frac{1}{2}(-6)$$

$$x_{MP} = 9 \qquad y_{MP} = -3$$

The correct choice is **(4)**.

**6.** The center and radius of the circle are needed to write its formula. The midpoint of the diameter gives the center:

$$x_{MP} = \frac{1}{2}(x_1 + x_2) \qquad y_{MP} = \frac{1}{2}(y_1 + y_2)$$

$$x_{MP} = \frac{1}{2}(4 + 8) \qquad y_{MP} = \frac{1}{2}(-1 + 3)$$

$$x_{MP} = \frac{1}{2}(12) \qquad y_{MP} = \frac{1}{2}(2)$$

$$x_{MP} = 6 \qquad y_{MP} = 1$$

The radius is the distance from the center point to either endpoint of the diameter. Apply the distance formula with points $(6, 1)$ and $(8, 3)$.

$$\begin{aligned}
\text{distance} &= \sqrt{(x_1 - x_2)^2 + (y_1 - y_2)^2} \\
&= \sqrt{(8 - 6)^2 + (3 - 1)^2} \\
&= \sqrt{(2)^2 + (2)^2} \\
&= \sqrt{8}
\end{aligned}$$

The radius of the circle is $\sqrt{8}$, and its center has coordinates $(6, 1)$. Substitute these values for $r$, $h$, and $k$ in the equation of a circle:

$$(x - h)^2 + (y - k)^2 = R^2$$
$$(x - 6)^2 + (y - 1)^2 = \sqrt{8}^2$$
$$(x - 6)^2 + (y - 1)^2 = 8$$

The correct choice is **(3)**.

**7.** Two segments are congruent if their lengths are equal, so apply the distance formula to determine the length of each segment.

$$d = \sqrt{(x_1 - x_2)^2 + (y_1 - y_2)^2}$$

$$AB = \sqrt{(-3-1)^2 + (6-4)^2} \qquad TU = \sqrt{(4-2)^2 + (1-5)^2}$$

$$= \sqrt{(-4)^2 + (2)^2} \qquad\qquad = \sqrt{(2)^2 + (-4)^2}$$

$$= \sqrt{16+4} \qquad\qquad\qquad = \sqrt{4+16}$$

$$= \sqrt{20} \qquad\qquad\qquad\quad = \sqrt{20}$$

$AB = TU$; therefore, the 2 segments are congruent.

**8.**
$$\frac{UW}{WV} = \frac{1}{5} = \frac{x - (-3)}{9 - x}$$

$$9 - x = 5(x + 3)$$

$$9 - x = 5x + 15$$

$$-6 = 6x$$

$$x = -1$$

Repeating for the $y$-coordinate:

$$\frac{UW}{WV} = \frac{1}{5} = \frac{y - 7}{1 - y}$$

$$1 - y = 5(y - 7)$$

The coordinates of W are $(-1, 6)$.

**9.** Line $m$ is the line of reflection that maps $A$ to $B$, so it must be the perpendicular bisector of $\overline{AB}$. To find the perpendicular bisector, calculate the midpoint and slope of $\overline{AB}$. Then write the equation of the line with the negative reciprocal slope that passes through the midpoint.

$$x_{MP} = \frac{1}{2}(x_1 + x_2), y_{MP} = \frac{1}{2}(y_1 + y_2)$$

$$x_{MP} = \frac{1}{2}(-2 + 6), y_{MP} = \frac{1}{2}(7 + 3)$$

$$x_{MP} = 2, \qquad\qquad y_{MP} = 5$$

$$\text{slope} = \frac{y_2 - y_1}{x_2 - x_1}$$

$$\text{slope}_{AB} = \frac{3 - 7}{6 - (-2)}$$

$$= -\frac{1}{2}$$

$\text{slope}_{\text{line } m} = 2$      $\perp$ lines have negative reciprocal slopes

$y - y_1 = m(x - x_1)$      point-slope equation of a line

$y - 5 = 2(x - 2)$ or $y = 2x + 1$      substitute the coordinates of the midpoint for $x_1$ and $y_1$, and 2 for $m$

**10.** Apply the completing the square procedure to rewrite the circle in $(x - h)^2 + (y - k)^2 = R^2$ form. Rewrite the equation with the variables on the left and constant on the right.

$$x^2 + 6x + y^2 - 12y + 25 = 0$$
$$x^2 + 6x + y^2 - 12y = -25$$

The constant needed to complete the square is $\left(\frac{1}{2}b\right)^2$, where $b$ is the coefficient of the linear $x$- and $y$-terms. For the $x$-terms, the necessary constant is $\left(\frac{1}{2}(6)\right)^2$, or 9. For the $y$-terms $\left(\frac{1}{2}(-12)\right)^2$, or 36, is needed. Add the required constants to each side of the equation.

$$x^2 + 6x + 9 + y^2 - 12y + 36 = -25 + 9 + 36$$

$(x + 3)^2 + (y - 6)^2 = 20$      factor the $x$-terms and the $y$-terms

The center has coordinates $(-3, 6)$ and the radius is $\sqrt{20}$.

**11.** A point that lies on a curve will satisfy the equation of the curve, so the strategy is to write the equation of the circle. Then substitute the coordinates of point $A$ for $x$ and $y$ in the equation and check if the equation is balanced.

$$(x-h)^2 + (y-k)^2 = R^2 \quad \text{equation of a circle}$$
$$(x-4)^2 + (y+5)^2 = \sqrt{65}^2 \quad \text{substitute } h = 4, k = -5, \text{ and } R = \sqrt{65}$$
$$(8-4)^2 + (2+5)^2 = \sqrt{65}^2 \quad \text{substitute coordinates of } A \text{ for } x \text{ and } y$$
$$4^2 + 7^2 = \sqrt{65}^2$$
$$16 + 49 = 65$$
$$65 = 65$$

The equation is balanced, so the point $A(8, 2)$ does lie on circle $P$.

**12.** Opposite sides of a parallelogram are parallel, so side $\overline{CD}$ is parallel to side $\overline{AB}$ and must have the same slope.
Start by finding the slope of $\overline{AB}$.

$$\text{slope } \overline{AB} = \frac{y_2 - y_1}{x_2 - x_1}$$

$$= \frac{1 - (-1)}{5 - 2}$$

$$= \frac{2}{3}$$

The desired line also passes through point $D(3, 4)$; $x = 3$ and $y = 4$ must satisfy the equation of $\overline{CD}$. Apply the point-slope equation of a line with slope $= \frac{2}{3}$, which represents the slope of $\overline{AB}$ using $x_1 = 3$ and $y_1 = 4$:

$$y - y_1 = m(x - x_1)$$
$$y - 4 = \frac{2}{3}(x - 3)$$

or, in slope-intercept form $y = \frac{2}{3}x + 2$.

**13.** We can estimate the area by filling in the ellipse with 5 rectangles. Triangles could have been used to get a more accurate estimate of the area.

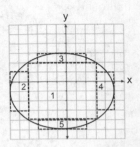

The areas of the rectangles are

rectangle 1 area = 7 · 6 = 42
rectangle 2 area = 4 · 2 = 8
rectangle 3 area = 5 · 1 = 5
rectangle 4 area = 4 · 2 = 8
rectangle 5 area = 5 · 1 = 5

The approximate area of the ellipse is 42 + 8 + 5 + 8 + 5 = 68 square units.

**14.** The two circles are the largest that will fit inside the region, and the smallest that will fit outside the region.

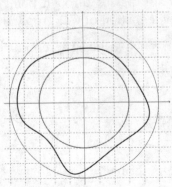

The inner circle has a radius of 3 cm and the outer circle has a radius of 5 cm. The estimate is the average of the areas of the two circles.

$$A_{estimate} = \frac{1}{2}\left(\pi R^2_{inner} + \pi R^2_{outer}\right)$$

$$= \frac{1}{2}\left(\pi \cdot 3^2 + \pi \cdot 5^2\right)$$

$$= \frac{1}{2}(106.8141)$$

$$\approx 53.4 \text{ cm}^2$$

**15.** a) To find the equation of a perpendicular bisector, the slope and midpoint of $\overline{SK}$ are needed.

$$\text{slope } \overline{SK} = \frac{y_2 - y_1}{x_2 - x_1}$$

$$= \frac{0-2}{6-0}$$

$$= -\frac{1}{3}$$

midpoint $\overline{SK}$ is $\dfrac{x_1 + x_2}{2}, \dfrac{y_1 + y_2}{2}$

$$\left(\frac{0+6}{2}, \frac{2+0}{2}\right)$$

$(3, 1)$

Perpendicular lines have negative reciprocal slopes, so the slope of the perpendicular bisector is 3, and the line must pass through the midpoint $(3, 1)$. Substitute these values in the point-slope equation of a line:

$$y - y_1 = m(x - x_1)$$
$$y - 1 = 3(x - 3)$$

or $y = 3x - 8$ in slope-intercept form.

b) An altitude will pass through the opposite vertex, so check if the coordinates of point $Y(4, 4)$ satisfy the equation of the perpendicular bisector.

$$y - 1 = 3(x - 3),$$
$$4 - 1 = 3(4 - 3)$$
$$3 = 3(1)$$
$$3 = 3$$

The equation is balanced, so the perpendicular bisector passes through the opposite vertex and is also an altitude.

## 3.7 SIMILAR FIGURES

### PROPERTIES AND DEFINITION OF SIMILAR FIGURES

Two figures are similar if there exists a sequence of similarity transformations that maps one figure onto another. Remember, the sequence of similarity transformations often includes a dilation.

Similar figures have the following properties:

- Same shape but possibly different sizes
- All corresponding lengths proportional
- All corresponding angles congruent

The ratio of proportionality is called the *scale factor*. A similarity statement is written using the symbol ~. In the accompanying figure, $\triangle ABC \sim \triangle DEF$, from the similarity statement, we have the following relationships:

$$\angle A \cong \angle D \quad \angle B \cong \angle E \quad \angle C \cong \angle F$$

$$\frac{AB}{DE} = \frac{BC}{EF} = \frac{AC}{DF} = \frac{1}{3}$$

 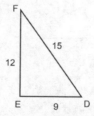

## PROVING TRIANGLES SIMILAR USING THE SIMILARITY POSTULATES

- Two triangles can be proven similar using the following postulates:

    **AA** (angle–angle)
    **SAS** (side–angle–side)
    **SSS** (side–side–side)

    Unlike congruence, pairs of corresponding sides must be in the same ratio.

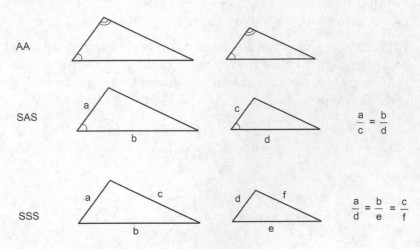

AA

SAS    $\dfrac{a}{c} = \dfrac{b}{d}$

SSS    $\dfrac{a}{d} = \dfrac{b}{e} = \dfrac{c}{f}$

## PERIMETER, AREA, AND VOLUME IN SIMILAR FIGURES

Perimeter, area, and volume in similar figures and solids are related to the ratio of corresponding sides.

| $\dfrac{\text{Perimeter}_A}{\text{Perimeter}_B}$ | $\dfrac{\text{Area}_A}{\text{Area}_B}$ | $\dfrac{\text{Volume}_A}{\text{Volume}_B}$ |
|:---:|:---:|:---:|
| scale factor | (scale factor)$^2$ | (scale factor)$^3$ |

# SIMILARITY RELATIONSHIPS IN TRIANGLES

*Segment Parallel to a Side Theorem*
- A segment parallel to a side of a triangle forms a triangle similar to the original triangle.
- If a segment intersects two sides of a triangle such that a triangle similar to the original is formed, the segment is parallel to the third side of the original triangle.

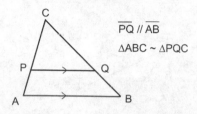

$\overline{PQ} \parallel \overline{AB}$

$\triangle ABC \sim \triangle PQC$

*Side Splitter Theorem*
A segment parallel to a side in a triangle divides the two sides it intersects proportionally.

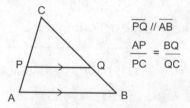

$\overline{PQ} \parallel \overline{AB}$

$\dfrac{AP}{PC} = \dfrac{BQ}{QC}$

---

***Centroid Theorem***
The centroid of a triangle divides each median in a 1 : 2 ratio, with the longer segment having a vertex as one of its endpoints.

---

In the accompanying figure, $D$ is the midpoint of $\overline{AB}$, and $E$ is the midpoint of $\overline{BC}$, making $\overline{AE}$ and $\overline{CD}$ medians. Their point of intersection, $G$, is the centroid and divides the medians in a 1 : 2 ratio. $\dfrac{EG}{AG} = \dfrac{1}{2}$ and $\dfrac{DG}{CG} = \dfrac{1}{2}$

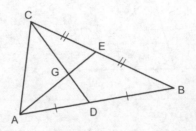

---

***Midsegment Theorem***
A segment joining the midpoints of two sides of a triangle (a midsegment) is parallel to the opposite side, and its length is equal to $\dfrac{1}{2}$ the length of the opposite side.

---

In the accompanying figure, $E$ is the midpoint of $\overline{AB}$, and $F$ is the midpoint of $\overline{AC}$. $\overline{EF}$ // $\overline{BC}$ and $EF = \dfrac{1}{2} BC$.

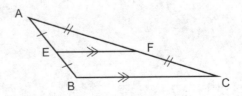

> **Altitude to the Hypotenuse of a Right Triangle Theorem**
> The altitude to the hypotenuse of a right triangle forms two triangles that are similar to the original triangle.

In the accompanying figure, $\overline{CD}$ is an altitude to hypotenuse $\overline{AB}$, $\triangle BDC \sim \triangle CDA \sim \triangle BCA$.

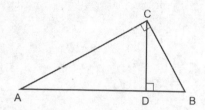

Two useful strategies for dealing with these overlapping similar triangles is to sketch the triangles separately or to make a table with sides classified as the "leg 1," "leg 2," and "hypotenuse." The altitude will be leg 1 of one of the interior triangles, and leg 2 of the other.

| | **Leg 1** | **Leg 2** | **Hypotenuse** |
|---|---|---|---|
| Small △ | $\overline{BD}$ | $\overline{CD}$ | $\overline{BC}$ |
| Medium △ | $\overline{CD}$ | $\overline{AD}$ | $\overline{AC}$ |
| Large △ | $\overline{BC}$ | $\overline{AC}$ | $\overline{AB}$ |

# Practice Exercises

1. In the accompanying figure, $\triangle PQR \sim \triangle TSR$ and points $P$, $R$, and $T$ are collinear. Which of the following transformations will map $\triangle TSR$ to $\triangle PQR$?

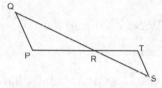

(1) reflection over line $\overline{TP}$ followed by a dilation with a scale factor of $\dfrac{QR}{TR}$ centered at $Q$

(2) translation by vector $\overrightarrow{TR}$ followed by a dilation with a scale factor of $\dfrac{PQ}{ST}$ centered at $R$

(3) rotation of $180°$ about point $R$ followed by a dilation with a scale factor of $\dfrac{PQ}{ST}$ centered at $R$

(4) rotation of $180°$ about point $R$ followed by a dilation with a scale factor of $\dfrac{QR}{TR}$ centered at $R$

2. *ABCDE* and *VWXYZ* are similar pentagons. If *BC* = 4 and *WX* = 6, what is the ratio of the area of *ABCDE* to the area of *VWXYZ*?

   (1) $\dfrac{4}{9}$          (3) $\dfrac{2}{5}$

   (2) $\dfrac{2}{3}$          (4) $\sqrt{\dfrac{2}{3}}$

3. If $\triangle XYZ \sim \triangle XVW$, which of the following transformations will map $\triangle XVW$ to $\triangle XYZ$?

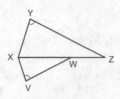

   (1) reflect $\triangle XVW$ over $\overline{XW}$ and then dilate it with a center at $X$ and scale factor of $\dfrac{XW}{WZ}$

   (2) reflect $\triangle XVW$ over $\overline{XW}$ and then dilate it with a center at $X$ and scale factor of $\dfrac{XZ}{XW}$

   (3) rotate $\triangle XVW$ 180° about point $X$ and then dilate it with a center at $X$ and scale factor of $\dfrac{XW}{WZ}$

   (4) rotate $\triangle XVW$ 180° about point $X$ and then dilate it with a center at $X$ and scale factor of $\dfrac{XZ}{XW}$

**4.** Which of the following triangles can be proven to be similar?

(1)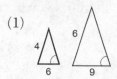

(3)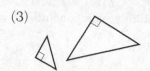

(2)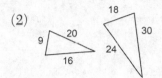

(4)

**5.** Given triangles *FOG* and *BAT* with $\angle G \cong \angle T$, $\angle O \cong \angle B$, $FG = 5x$, $FO = 42$, $AB = 14$, and $AT = x + 12$, find the length *FG*.
(1) 18                    (3) 45
(2) 30                    (4) 90

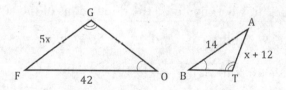

**6.** $\overline{UV}$ is parallel to side $\overline{ST}$ of $\triangle RST$. If $RU = x - 6$, $US = x$, $RV = x$, and $VT = 2x - 9$, what is the value of $x$?

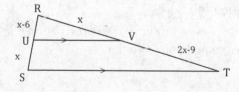

(1) 3                    (3) 18
(2) 11                    (4) 39

7. In the figure below, $D$ lies on $\overline{AB}$ and $E$ lies on $\overline{AC}$. If $AD = 3$, $BD = x + 3$, $DE = 5$, and $BC = 2x + 2$, find the value of $x$ that would let you conclude that $\overline{DE} \parallel \overline{BC}$.

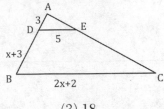

(1) 2                (3) 18

(2) 9                (4) 24

8. Mollie wants to rearrange the furniture in her living room. She measures the rectangular room to be 16 ft by 12 ft and makes a scaled drawing of the room that measures 10 in by 7.5 in. She wants to make scaled paper cutouts to represent her furniture so she can determine her favorite arrangement without having to move the furniture. If her sofa is rectangular and measures 6 ft by 2 ft, what should be the dimensions of the cutout?

(1) $1\dfrac{1}{3}$ in $\times$ 1 in        (3) $3\dfrac{3}{4}$ in $\times$ $1\dfrac{1}{4}$ in

(2) $2\dfrac{3}{4}$ in $\times$ 1 in        (4) 9.6 in $\times$ 3.2 in

9. Given: $\overline{LSI}$ and $\overline{FSP}$ intersecting at $S$, $\overline{FL} \parallel \overline{IP}$, $IP = 4$ and $FL = 6$

(1) Explain why the two triangles formed are similar.

(2) Find the ratio $\dfrac{PS}{FS}$.

(3) State a specific similarity transformation that will map the smaller triangle onto the larger.

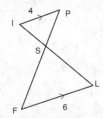

**10.** In $\triangle VTP$, $F$ is the midpoint of $\overline{VP}$, $G$ is the midpoint of $\overline{PT}$, and $\overline{VG}$ and $\overline{FT}$ intersect at $K$. If $VK = 5x + 5$ and $KG = 2x + 8$, find the length $VG$.

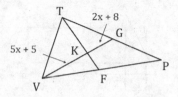

**11.** The midpoints of $\overline{PG}$, $\overline{PI}$, and $\overline{GI}$ are $C$, $O$, and $W$, respectively. If $OW = 7$, $CW = 5$, and $OC = 6$, find the perimeter of $\triangle PIG$.

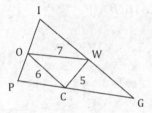

**12.** In the figure below, $\angle ABC$ and $\angle BDC$ are right angles, $CB = 6$, and $AB = 12$. Find the length of $\overline{BD}$. Express your answer in simplest radical form.

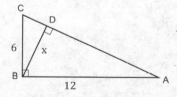

**13.** Rectangle *AEFG* is the image of rectangle *ABCD* after the transformation clockwise $R_{90°, A} \circ D_{\frac{2}{3}, A}$. If *GA* = 8 and *DE* = 12, what is the perimeter of *ABCD*?

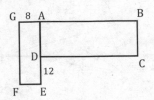

**14.** *ABCDEF* and *ARSTUV* are both regular hexagons and *R* is the midpoint of $\overline{AB}$. Name a sequence of transformations that will map *ARSTUV* to *ABCDEF*.

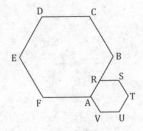

**15.** Given: $\overline{RUT}$, $\overline{SVT}$, and $\overline{UV} \parallel \overline{RS}$

Prove: $\dfrac{UR}{TU} = \dfrac{VS}{TV}$

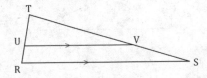

**16.** Given: $\overline{ABD}$, $\angle C \cong \angle E$, $\angle DEB \cong \angle EBC$
Prove: $\triangle ABC \sim \triangle BDE$

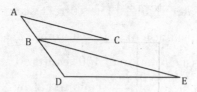

**17.** Prove the Pythagorean theorem.
Given: right triangle $\triangle BCA$ with a right angle at $C$, altitude $\overline{CD}$ drawn to hypotenuse $\overline{AB}$

Prove: $(AC)^2 + (BC)^2 = (AB)^2$

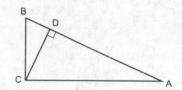

**Solutions**

**1.** We want to map the smaller triangle to the larger, so the scale factor will be greater than 1. The correct scale factor among the choices is $\frac{PQ}{ST}$. A rotation of 180° will rotate $\triangle TSR$ to $\triangle PQR$. The correct sequence is: a rotation of 180° centered at $R$, followed by a dilation by a scale factor of $\frac{PQ}{ST}$ centered at $R$. Note that the order of the sequence does not matter in this situation.

The correct choice is **(3)**.

**2.** The ratio of the areas of similar figures is equal to the ratio of lengths of corresponding sides squared. $\overline{BC}$ and $\overline{WX}$ are corresponding sides, and are in a ratio of $\frac{4}{6}$, or $\frac{2}{3}$.

$$\text{ratio of areas} = \text{ratio of sides}^2$$
$$= \left(\frac{2}{3}\right)^2$$
$$= \frac{4}{9}$$

The correct choice is **(1)**.

**3.** A reflection over $\overline{XW}$ will map $\overline{XV}$ onto $\overline{XZ}$. Then a dilation centered about $X$ with a scale factor of $\frac{XZ}{XW}$ will match the sizes. The correct choice is **(2)**.

**4.** The triangle similarity postulates are **AA**, **SAS**, and **SSS**, where each S represents the *ratio* of a pair of corresponding sides. Examine each of the choices.

Choice 1: The two pairs of corresponding sides are in a 2:3 ratio, but the congruent angle is not the included angle.

Choice 2: Corresponding sides are not in the same ratio, $\frac{9}{18} \neq \frac{16}{24}$.

Choice 3: Lengths of the sides are not given.

Choice 4: 2 pairs of corresponding sides are in the same ratio, $\frac{6}{8} = \frac{9}{12}$, and the congruent vertical angles are the included angles. The triangles are similar by SAS.

The correct choice is **(4)**.

**5.** The two triangles are similar by AA, and corresponding sides of similar triangles are in the same ratio. The similarity statement can be written as $\triangle FOG \sim \triangle ABT$, so the following proportion can be written:

$$\frac{FG}{AT} = \frac{FO}{AB}$$

$$\frac{5x}{x+12} = \frac{42}{14}$$

$$14(5x) = 42(x + 12) \qquad \text{cross products are equal}$$

$$70x = 42x + 504$$
$$28x = 504$$
$$x = 18 \qquad \text{substitute } x = 18$$
$$FG = 5(18)$$
$$= 90$$

The correct choice is **(4)**.

**6.** When a segment is drawn parallel to a side of a triangle, the two triangles are similar, and the side splitter theorem states that the sides are divided proportionally.

$$\frac{RU}{US} = \frac{RV}{VT} \qquad \text{side splitter theorem}$$

$$\frac{x-6}{x} = \frac{x}{2x-9}$$

$$(2x - 9)(x - 6) = x^2 \qquad \text{cross products are equal}$$
$$2x^2 - 21x + 54 = x^2$$
$$x^2 - 21x + 54 = 0 \qquad \text{solve by moving all terms to one side}$$
and factoring
$$(x - 18)(x - 3) = 0 \qquad \text{factor the trinomial}$$
$$x - 18 = 0 \qquad x - 3 = 0 \qquad \text{zero product theorem}$$
$$x = 18 \qquad x = 3$$

The two solutions of the quadratic are *possible* solutions to the problem. Eliminate any results that lead to negative lengths or any other impossible situations. The solution $x = 3$ would result in a negative length for $RU$ and $VT$, so we eliminate this result, and the value of $x$ is 18.

The correct choice is **(3)**.

**7.** From the original figure, $\overline{DE} \parallel \overline{BC}$ when $\angle CBA \cong \angle EDA$ (corresponding angles are congruent). The triangles $\triangle ADE$ and $\triangle ABC$ are similar by AA, and corresponding sides will be proportional. Sketch the two triangles separately to help identify the corresponding sides. Note that $AB = AD + DB$.

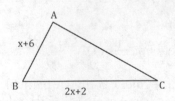

$$\frac{AD}{AB} = \frac{DE}{BC} \qquad \text{corresponding sides are proportional}$$
$$\frac{3}{x+6} = \frac{5}{2x+2}$$

$$3(2x + 2) = 5(x + 6) \qquad \text{cross products are equal}$$
$$6x + 6 = 5x + 30$$
$$x = 24$$

The correct choice is **(4)**.

**8.** First convert feet to inches so all units are consistent.

Length of room $\qquad$ $16\,\text{ft} \times \dfrac{12\,\text{in}}{\text{ft}} = 192\,\text{in}$

Length of sofa $\qquad$ $6\,\text{ft} \times \dfrac{12\,\text{in}}{\text{ft}} = 72\,\text{in}$

Width of sofa $\qquad$ $2\,\text{ft} \times \dfrac{12\,\text{in}}{\text{ft}} = 24\,\text{in}$

The scaled drawing is similar to the actual room, so the ratio of corresponding sides is

$$\frac{\text{length of drawing}}{\text{length of room}} = \frac{10}{192}$$

Apply this scale factor to the length and width of the sofa to determine the dimensions of the cutout of the sofa.

$$\frac{\text{length}_{\text{cutout}}}{\text{length}_{\text{actual}}} = \frac{10}{192} \qquad \frac{\text{width}_{\text{cutout}}}{\text{width}_{\text{actual}}} = \frac{10}{192}$$

$$\frac{\text{length}_{\text{cutout}}}{72\,\text{in}} = \frac{10}{192} \qquad \frac{\text{width}_{\text{cutout}}}{24\,\text{in}} = \frac{10}{192}$$

$$\text{length}_{\text{cutout}} = \frac{72 \cdot 10}{192} \qquad \text{width}_{\text{cutout}} = \frac{24 \cdot 10}{192}$$

$$\text{length}_{\text{cutout}} = 3.75\,\text{in} \qquad \text{width}_{\text{cutout}} = 1.25\,\text{in}$$

Converting decimals to fractions, the dimensions of the cutout are $3\dfrac{3}{4} \times 1\dfrac{1}{4}$.

The correct choice is **(3)**.

**9.** a) Given $\overline{FL} \parallel \overline{IP}$, we know $\angle F \cong \angle P$ and $\angle L \cong \angle I$ because they are the alternate interior angles formed by parallel lines and a transversal. Therefore, $\triangle SIP \sim \triangle SLF$ by AA. (Note the correct order of vertices in the similarity statement.)

b) Corresponding parts are in the same ratio, so

$$\frac{PS}{FS} = \frac{IP}{LF}$$

$$= \frac{4}{6}$$

$$= \frac{2}{3}$$

c) We have to use the reciprocal of the ratio in part (b) because we want to enlarge the smaller triangle and not reduce it. A rotation of $180°$ about point $S$ will align $\overline{SP}$ with $\overline{FS}$ and $\overline{IS}$ with $\overline{SL}$. A dilation with center at $S$ and a scale factor of $\frac{3}{2}$ will then map $\triangle SIP$ to $\triangle SLF$.

**10.** $\overline{VG}$ and $\overline{TF}$ are medians that intersect at centroid $K$. The centroid divides $\overline{VG}$ in a 1 : 2 ratio, so

$$\begin{aligned}
VK &= 2KG \\
5x + 5 &= 2(2x + 8) \\
5x + 5 &= 4x + 16 \\
5x &= 4x + 11 \\
x &= 11
\end{aligned}$$

$VG$ is equal to the sum of $VK$ and $KG$, which can be evaluated for $x = 11$.

$$\begin{aligned}
VG &= VK + KG \\
&= 5x + 5 + 2x + 8 \\
&= 7x + 13 \\
&= 7(11) + 13 \\
&= 90
\end{aligned}$$

**11.** $\overline{CO}$, $\overline{OW}$, and $\overline{WC}$ are all midsegments, and their lengths are $\frac{1}{2}$ the length of the opposite side of $\triangle PIG$. Therefore,

$$\begin{aligned}
PI &= 2 \cdot WC \\
&= 2 \cdot 5 \\
&= 10
\end{aligned}$$

$$IG = 2 \cdot OC$$
$$= 2 \cdot 6$$
$$= 12$$

$$GP = 2 \cdot WO$$
$$= 2 \cdot 7$$
$$= 14$$

The perimeter of $\triangle PIG = PI + IG + GP$
$$= 10 + 12 + 14$$
$$= 36$$

**12.** An altitude to the hypotenuse of a right triangle forms 3 similar triangles. The three similar triangles are $\triangle CDB$, $\triangle BDA$, and $\triangle CBA$. Fill in the table of side lengths, and look for two pairs of corresponding sides that can be used to write a proportion involving the unknown length $x$.

|  | **Leg 1** | **Leg 2** | **Hypotenuse** |
|---|---|---|---|
| Small $\triangle CDB$ |  | $x$ | 6 |
| Medium $\triangle BDA$ | $x$ |  | 12 |
| Large $\triangle CBA$ | 6 | 12 |  |

We don't have a full set of four corresponding sides, so we need to use the Pythagorean theorem on the large triangle to find the hypotenuse $AC$.

$$BC^2 + AB^2 = AC^2$$
$$6^2 + 12^2 = AC^2$$
$$AC^2 = 180$$
$$AC = \sqrt{180}$$
$$= 6\sqrt{5}$$

Now the table indicates two pairs of corresponding parts can be used to write a proportion.

|  | Leg 1 | Leg 2 | Hypotenuse |
|---|---|---|---|
| Small $\triangle CDB$ |  | $x$ | 6 |
| Medium $\triangle BDA$ | $x$ |  | 12 |
| Large $\triangle CBA$ | 6 | 12 | $6\sqrt{5}$ |

$$\frac{x}{12} = \frac{6}{6\sqrt{5}}$$

$$6x\sqrt{5} = 72 \qquad \text{cross products are equal}$$

$$x = \frac{72}{6\sqrt{5}}$$

$$= \frac{12}{\sqrt{5}}$$

$$= \frac{12}{\sqrt{5}} \cdot \frac{\sqrt{5}}{\sqrt{5}} \qquad \text{rationalize the denominator}$$

$$= \frac{12\sqrt{5}}{5}$$

**13.** The dilation tells us that the ratio of corresponding parts between rectangles $AEFG$ and $ABCD$ is 2:3. $GA$ and $AD$ are corresponding parts, so the following proportion can be written:

$$\frac{GA}{AD} = \frac{2}{3}$$

$$\frac{8}{AD} = \frac{2}{3}$$

$$2AD = 24 \qquad \text{cross products are equal}$$
$$AD = 12$$
$$AE = AD + DE \quad \text{segment addition}$$
$$AE = 12 + 12$$
$$= 24$$

Now we can find $AB$ by writing a proportion involving corresponding sides $AE$ and $AB$.

$$\frac{AE}{AB} = \frac{2}{3}$$

$$3AE = 2AB \qquad \text{cross products are equal}$$
$$3(24) = 2AB$$
$$72 = 2AB$$
$$AB = 36$$

Since $ABCD$ is a rectangle, opposite sides are congruent, and the perimeter is equal to

$$\text{perimeter}_{ABCD} = 2AB + 2BC$$
$$= 2(36) + 2(12)$$
$$= 72 + 24$$
$$= 96$$

**14.** A dilation and a reflection are required to map $ARSTUV$ to $ABCDEF$. Since $R$ is a midpoint, we know $AR = \frac{1}{2}AB$, and a dilation with a scale factor of 2 centered at $A$ is needed to map $\overline{AR}$ to $\overline{AB}$. The second transformation needed is a reflection over line $\overleftrightarrow{AB}$.

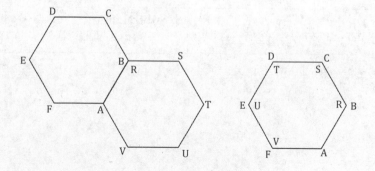

The transformation is a dilation centered at $A$ with a scale factor of 2, followed by a reflection over $\overline{AB}$. Alternatively, the order could be reversed.

**15.** The strategy is to prove $\triangle TUV \sim \triangle TRS$, which leads to the proportion of corresponding sides $\dfrac{TR}{TU} = \dfrac{TS}{TV}$. Algebraic rearranging gives the final result.

| Statement | Reason |
|---|---|
| 1. $\overline{RUT}$, $\overline{SVT}$, and $\overline{UV} \parallel \overline{RS}$ | 1. Given |
| 2. $\angle TUV \cong \angle R$<br>   and $\angle TVU \cong \angle S$ | 2. Corresponding angles formed by parallel lines are congruent |
| 3. $\triangle TUV \sim \triangle TRS$ | 3. AA |
| 4. $\dfrac{TR}{TU} = \dfrac{TS}{TV}$ | 4. Corresponding sides in similar triangles are proportional |
| 5. $TR = TU + UR$<br>   $TS = TV + VS$ | 5. Partition |
| 6. $\dfrac{TU + UR}{TU} = \dfrac{TV + VS}{TV}$ | 6. Substitution |
| 7. $1 + \dfrac{UR}{TU} = 1 + \dfrac{VS}{TV}$ | 7. Division |
| 8. $\dfrac{UR}{TU} = \dfrac{VS}{TV}$ | 8. Subtraction |

**16.** The strategy is to show that $\overline{BC}$ // $\overline{DE}$ and then to use the corresponding angle relationship to show $\angle ABC \cong \angle BDE$.

| Statement | Reason |
|---|---|
| 1. $\overline{ABD}$, $\angle C \cong \angle E$, $\angle DEB \cong \angle EBC$ | 1. Given |
| 2. $\overline{BC}$ // $\overline{DE}$ | 2. Two lines are parallel if the alternate interior angles formed by a transversal are congruent |
| 3. $\angle ABC \cong \angle BDE$ | 3. The corresponding angles formed by two parallel lines are congruent |
| 4. $\triangle ABC \sim \triangle BDE$ | 4. AA |

**17.** The strategy is to use the theorem that an altitude to the hypotenuse of a right triangle forms three similar triangles. Two proportions from these triangles can be written and combine to yield the Pythagorean theorem.

| Statement | Reason |
|---|---|
| 1. Right triangle $\triangle ABC$ with a right angle at $C$, altitude $\overline{CD}$ drawn to hypotenuse $\overline{AC}$ | 1. Given |
| 2. $\triangle BDC \sim \triangle CDA \sim \triangle BCA$ | 2. The altitude to the hypotenuse of a right triangle forms 3 similar right triangles |
| 3. $\dfrac{AB}{AC} = \dfrac{AC}{AD}$, $\dfrac{AB}{BC} = \dfrac{BC}{BD}$ | 3. Corresponding sides in similar triangles are proportional |
| 4. $(AC)^2 = AB \cdot AD$, $(BC)^2 = AB \cdot BD$ | 4. The cross products of a proportion are equal |
| 5. $(AC)^2 + (BC)^2 = AB \cdot AD + AB \cdot BD$ | 5. Addition property |
| 6. $(AC)^2 + (BC)^2 = AB(AD + BD)$ | 6. Factor |
| 7. $(AC)^2 + (BC)^2 = AB(AB)$ | 7. Partition property |
| 8. $(AC)^2 + (BC)^2 = (AB)^2$ | 8. Simplify |

## 3.8. TRIGONOMETRY

### DEFINITION OF SIDES IN A RIGHT TRIANGLE RELATIVE TO AN ANGLE

- Hypotenuse—the side across from the right angle
- Opposite—the side across from the specified acute angle
- Adjacent—the side included between the right angle and the specified angle

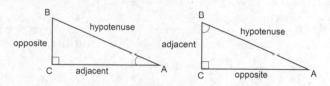

a) Opposite, adjacent, and hypotenuse relative to ∠A

b) Opposite, adjacent, and hypotenuse relative to ∠B

### TRIGONOMETRIC RATIOS

| Ratio | Abbreviation | Definition |
|---|---|---|
| Sine | Sin | $\dfrac{\text{opposite}}{\text{hypotenuse}}$ |
| Cosine | Cos | $\dfrac{\text{adjacent}}{\text{hypotenuse}}$ |
| Tangent | Tan | $\dfrac{\text{opposite}}{\text{adjacent}}$ |

The ratios sine, cosine, and tangent will have fixed values for a given acute angle in a right triangle. This is because any two right triangles with a pair of congruent acute angles are similar, and corresponding parts of similar triangles are in the same ratio. These three trigonometric ratios can be used to find a missing side of a triangle given one angle and side by setting up the correct proportion. Be sure your calculator is in degree mode if the angle is given in degrees.

**Example:**
$\triangle SOX$ has a right angle at $O$. The measure of angle $S = 72°$ and $XO = 12$. Find the length $OS$ to the nearest tenth.

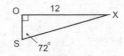

*Solution*:
Relative to $\angle S$, the given side and the unknown side are the opposite and adjacent, which suggest using the tangent ratio.

$$\tan(S) = \frac{\text{opposite}}{\text{adjacent}} = \frac{OX}{OS}$$

$$\tan(72°) = \frac{12}{OS}$$

$$OS = \frac{12}{\tan(72°)}$$

$$OS = \frac{12}{3.07768} = 3.89903$$

$$= 3.9$$

## INVERSE TRIGONOMETRIC RATIOS

The inverse of a function "undoes" the original function. For example, $f(x) = \sqrt{x}$ and $g(x) = x^2$ are inverse functions. If $x = 5$, then $f(g(5)) = \sqrt{5^2} = 5$. The trigonometric functions have the following inverse functions:

| | |
|---|---|
| $\sin(x)$ | $\arcsin(x)$ |
| $\cos(x)$ | $\arccos(x)$ |
| $\tan(x)$ | $\arctan(x)$ |

- These three inverse functions are often abbreviated $\sin^{-1}(x)$, $\cos^{-1}(x)$, and $\tan^{-1}(x)$. They can be used to find the measure of an angle given a ratio of any two sides in the triangle.

**Example:**
Using the accompanying figure, find the measure of angles $M$ and $N$.

*Solution:*

$$\cos(M) = \frac{3}{7}$$

$$m\angle M = \cos^{-1}\left(\frac{3}{7}\right)$$

$$= 64.6°$$

$$\sin(N) = \frac{3}{7}$$

$$m\angle N = \sin^{-1}\left(\frac{3}{7}\right)$$

$$= 25.4°$$

# COFUNCTIONS

### Cofunction Relationship
The sine of an angle is equal to the cosine of its complement: $\sin(A) = \cos(90 - A)$.

The cosine of an angle is equal to the sine of its complement: $\cos(A) = \sin(90 - A)$.

$$\sin(A) = \cos(B) = \cos(90 - A) = \frac{BC}{AB}$$

$$\sin(B) = \cos(A) = \sin(90 - A) = \frac{AC}{AB}$$

In other words, the cofunction relationship states that if a sine is equal to a cosine, then the two angles must sum to 90°.

**Example:**
Given $\sin(2x + 4) = \cos(3x + 21)$. Solve for $x$.

*Solution*:
The cofunctions are equal, so the angles must be complementary.

$$2x + 4 + 3x + 21 = 90$$
$$5x + 25 = 90$$
$$5x = 65$$
$$x = 13$$

## MODELING WITH TRIGONOMETRY

- *Angle of elevation*—The angle formed by the horizontal and the line directed upward to an object
- *Angle of depression*—The angle formed by the horizontal and the line directed downward to an object

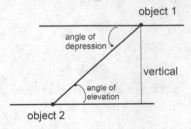

Strategy for trigonometry modeling problems:

- Start with a detailed sketch.
- Identify the triangles formed and the relevant trigonometric ratios.
- Consider parallel line and other relationships if you are still missing parts. Notice that the angle of elevation and angle of depression are congruent alternate interior angles.
- Consider working with more than one triangle, especially if the problem involves an object that moves from one position to another.

# Practice Exercises

1.  In $\triangle ABC$, m$\angle B$ = 90°, m$\angle C$ = 51°, and $BC$ = 13. What is the length $AB$?

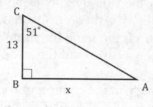

   (1) 16.1          (3) 18.2
   (2) 16.7          (4) 20.7

2.  $\triangle JPL$ is a right triangle with a right angle at $L$. If m$\angle J$ = 19° and $\overline{LJ}$ = 21, what is the length of $\overline{PJ}$?

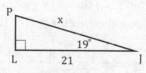

   (1) 21.5          (3) 61.0
   (2) 22.2          (4) 64.5

**3.** A 24 ft ladder is leaning against a wall. The base of the ladder is 5 ft from the house. What angle does the ladder make with the ground?
(1) 12°          (3) 67°
(2) 65°          (4) 78°

**4.** In right triangle $PQR$, $m\angle Q = 90°$. If $\sin(P) = 4x^2 + 5x + 2$ and $\cos(R) = 5x + 3$, what is the measure of $\angle P$?
(1) 5°           (3) 25°
(2) 8°           (4) 30°

**5.** John constructs three right triangles of different sizes, each having angles that measure 30°, 60°, and 90°. He then measures the side lengths of each triangle and calculates the value of $\sin(60°)$ from the side lengths using each of the triangles. Which of the following theorems would best explain why the value he calculates is the same for all three triangles?
(1) The sum of the measures of the interior angles of a triangle equals 180°.
(2) Corresponding parts of congruent triangles are congruent.
(3) Corresponding sides of similar triangles are proportional.
(4) The longest side of a triangle is always opposite the largest angle.

**6.** The side of Rayquan's house is 30 ft long. Rayquan is planning to install a gutter along the roof as shown. He wants the angle of depression to be 3° so the water will drain out at the end of the gutter. How long should the gutter be?

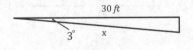

(1) 31 ft 7 in          (3) 30 ft 4 in
(2) 31 ft 2 in          (4) 30 ft 0.5 in

7. In $\triangle ABC$, m$\angle B$ = 90°, tan($A$) = $m$, and sin($C$) = $\dfrac{1}{p}$. What is the value of cos($C$)?

   (1) $\dfrac{m}{p}$             (3) $\dfrac{p}{m}$

   (2) $m \cdot p$            (4) $m + p$

8. Jack is watching the launch of a rocket from a viewing area 4,500 feet from the launch pad. At 15 seconds after launch, he measures a 67° angle of elevation from the ground to the rocket. What is the average speed of the rocket during the first 15 seconds of its flight? Assume the rocket travels upward, perpendicular to the ground, and give your answer to the nearest foot per second.

9. Rick is flying a kite in the park. He holds the spool of string 3 ft above the ground, and lets out 200 ft of string. The kite is initially flying with a 72° angle of elevation.
   a) What is the altitude of the kite?
   b) The wind shifts and the angle of elevation altitude decreases to 52°. How many feet does the kite drop?

10. Frank is standing at the top of a water tower and looks down at the town below. Frank measures an angle of depression of 52° to his house, and he knows the water tower is 120 ft tall. How far along the ground is the tower from his house? Round to the nearest foot.

**11.** A lighthouse sits on a 100 ft cliff. An observer in a boat in the harbor notes an angle of elevation to the base of the cliff of 28°, and an angle of elevation to the top of the lighthouse of 44°. What is the height of the lighthouse, rounded to the nearest foot?

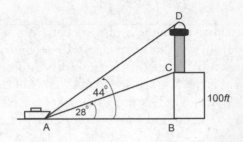

**Solutions**

**1.** Relative to $\angle C$, $\overline{AB}$ is the opposite and $\overline{BC}$ is the adjacent, so the tangent ratio applies.

$$\text{tangent} = \frac{\text{opposite}}{\text{adjacent}}$$

$$\tan(C) = \frac{AB}{BC}$$

$$\tan(51°) = \frac{x}{13}$$

$$\begin{aligned} x &= 13 \tan(51°) \\ &= 16.0536 \\ &= 16.1 \end{aligned}$$

The correct choice is (**1**).

**2.** Relative to $\angle J$, $\overline{LJ}$ is the adjacent and $\overline{JP}$ is the hypotenuse. The cosine ratio applies here.

$$\text{cosine} = \frac{\text{adjacent}}{\text{hypotenuse}}$$

$$\cos(J) = \frac{LJ}{JP}$$

$$\cos(19°) = \frac{21}{x}$$

$$x = \frac{21}{\cos(19°)}$$

$$= 22.2100$$

$$= 22.2$$

The correct choice is **(2)**.

**3.** The ladder, wall, and ground form a right triangle as shown.

Relative to the desired angle, the 5 ft distance is the adjacent and the 24 ft length is the hypotenuse. Adjacent and hypotenuse suggest using the cosine. Since we are looking for the angle, apply the inverse cosine function.

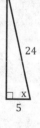

$$\cos(x) = \frac{\text{adjacent}}{\text{hypotenuse}}$$

$$\cos(x) = \frac{5}{24}$$

$$x = \cos^{-1}\left(\frac{5}{24}\right)$$

$$= 77.9753°$$

$$= 78°$$

The correct choice is **(4)**.

**4.** The sine of an angle and the cosine of its complement are always equal. Since $\triangle PQR$ is a right triangle with the right angle at $Q$, angles $P$ and $R$ must be complementary. Setting $\sin(P)$ and $\cos(R)$ equal, we get:

$$\sin(P) = \cos(R)$$
$$4x^2 + 5x + 2 = 5x + 3$$
$$4x^2 = 1$$
$$x^2 = \frac{1}{4}$$
$$x = \pm\frac{1}{2}$$

Substituting $x = -\frac{1}{2}$ into the expression for $\sin(P)$, we get

$$\sin(P) = 4x^2 + 5x + 2$$
$$= 4(-\frac{1}{2})^2 + 5\left(-\frac{1}{2}\right) + 2$$
$$= 1 - 2\frac{1}{2} + 2$$
$$= \frac{1}{2}$$

Now substitute $x = \frac{1}{2}$.

$$\sin(P) = 4\left(\frac{1}{2}\right)^2 + 5\left(\frac{1}{2}\right) + 2$$
$$= 1 + 2\frac{1}{2} + 2$$
$$= 5\frac{1}{2}$$

The sine function cannot be greater than 1, so $x = \frac{1}{2}$ is not a solution. Using the solution $x = -\frac{1}{2}$ and $\sin(P) = \frac{1}{2}$, find m$\angle P$ using the inverse sine function.

$$\sin(P) = \frac{1}{2}$$

$$m\angle P = \sin^{-1}\left(\frac{1}{2}\right)$$

$$m\angle P = 30°$$

The correct choice is **(4)**.

**5.** The three triangles are all similar by the angle–angle theorem. The sine ratio is equal to the $\frac{\text{opposite}}{\text{hypotenuse}}$. Since all pairs of corresponding sides in similar triangles are proportional, the ratio $\frac{\text{opposite}}{\text{hypotenuse}}$ will be equal to the same value in all three triangles.

The correct choice is **(3)**.

**6.** Relative to the 3° angle, the horizontal distance is the adjacent and the gutter is the hypotenuse, so we can apply the cosine ratio.

$$\cos(3°) = \frac{\text{adjacent}}{\text{hypotenuse}}$$

$$\cos(3°) = \frac{30 \text{ ft}}{x}$$

$$x = \frac{30 \text{ ft}}{\cos(3°)}$$

$$= 30.04117 \text{ ft}$$

To convert to feet and inches, multiply the decimal part of the number by 12.

$$0.04117 \,(12) = 0.49405 \text{ in}$$

The gutter length should be 30 ft 0.49405 in. Rounded to the nearest 0.1 inch, the length is 30 ft 0.5 in.

The correct choice is **(4)**.

**7.** Making a sketch of $\triangle ABC$, we can fill in the lengths of $AB$, $BC$, and $AC$ as shown in the figure.

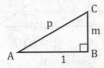

The tangent is equal to the $\dfrac{\text{opposite}}{\text{adjacent}}$, so

$$\tan(A) = m$$

$$\frac{BC}{AB} = \frac{m}{1}$$

Since we are only concerned with ratios, we can assume $BC = m$ and $AB = 1$.

The second ratio states

$$\sin(C) = \frac{1}{p}$$

$$\frac{AB}{AC} = \frac{1}{p}$$

Since we assumed $AB = 1$, we can assume $AC = p$. Cos$(C)$ can now be calculated.

$$\cos(C) = \frac{\text{adjacent}}{\text{hypotenuse}}$$

$$= \frac{BC}{AC}$$

$$= \frac{m}{p}$$

The correct choice is **(1)**.

**8.** The path of the rocket traveling upward forms a right triangle with the ground. The distance to the rocket is the adjacent, and the height of the rocket is the opposite, so the tangent ratio can be applied.

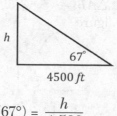

$$4500\,ft$$

$$\tan(67°) = \frac{h}{4,500}$$

$$h = 4,500\tan(67°)$$

$$= 4,500\,(2.3558)$$

$$= 10,601.33 \text{ ft}$$

To find the speed, apply the relationship rate $= \dfrac{\text{distance}}{\text{time}}$.

$$\text{speed} = \frac{10,601.33 \text{ ft}}{15 \text{ sec}}$$

$$= 706.73 \text{ ft/sec}$$

$$= 707 \text{ ft/sec}$$

**9.** First make a sketch. A right triangle formed by the kite string, the altitude measured from the spool ($x$), and the horizontal distance to the kite. The total altitude will be ($x + 3$) ft. To find how many feet the kite drops, calculate the altitude at the two angles of elevation, and then find the difference.

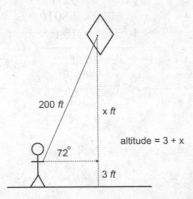

a) Relative to the angle of elevation, the 200 ft kite string is the hypotenuse, and the altitude measured from the spool is opposite. The opposite and hypotenuse suggest the sine ratio. Find $x$ using the following:

$$\sin(72°) = \frac{\text{opposite}}{\text{hypotenuse}}$$

$$\sin(72°) = \frac{x}{200}$$

$$x = 200 \sin(72°)$$

$$x = 200 \,(0.951056)$$

$$= 190.2112 \text{ ft}$$

Next, calculate the altitude from $x$.

$$\text{Altitude} = 3 \text{ ft} + x \text{ ft}$$

$$= 3 \text{ ft} + 190.21123 \text{ ft}$$

$$= 193.21123 \text{ ft}$$

$$= 193 \text{ ft}$$

b) We can use the same equations to calculate the new altitude if the angle of elevation decreases to 52°.

$$\sin(52°) = \frac{x}{200}$$
$$x = 200 \sin(52°)$$
$$x = 200 \, (0.788010)$$
$$= 157.60215 \text{ ft}$$

$$\text{Altitude} = 3 \text{ ft} + x \text{ ft}$$
$$= 3 + 157.60215 \text{ ft}$$
$$= 160.60215 \text{ ft}$$

The change in altitude is

$$193.21123 \text{ ft} - 160.60215 \text{ ft}$$

$$= 32.6090 \text{ ft}$$
$$= 33 \text{ ft}$$

The kite drops 33 ft.

**10.** Start with a sketch, and fill in the appropriate dimensions, horizontals, and verticals.

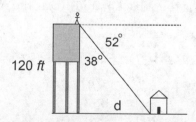

The angle of depression is 52°. It is easier in the problem to find the angle complementary to the angle of depression and work with that triangle. The complementary angle is 38°. The height of the tower is adjacent to the 38° angle, and the distance along the ground is opposite from the 38° angle, so we use the tangent ratio.

$$\tan(38°) = \frac{d}{120 \text{ ft}}$$
$$d = 120 \text{ ft} \cdot \tan(38°)$$
$$d = 93.75 \text{ ft}$$

The house is 94 ft from the water tower.

**11.** First find length $AB$ using $\triangle ABC$; then find length $BD$ using $\triangle ABD$.

In $\triangle ABC$, $BC$ is opposite $\angle A$, and $AB$ is the adjacent, so we can use the tangent.

$$\tan = \frac{\text{opposite}}{\text{adjacent}}$$

$$\tan(\angle BAC) = \frac{BC}{AB}$$

$$\tan(28°) = \frac{100}{AB}$$

$$AB = \frac{100}{\tan(28°)}$$

$$AB = 188.0726$$

In $\triangle ABD$, $AB$ is the adjacent and $BD$ is the opposite, so we use the tangent again.

$$\tan(\angle BAD) = \frac{BD}{AD}$$

$$\tan(44°) = \frac{BD}{188.0726}$$

$$BD = 188.0726 \tan(44°)$$

$$BD = 181.6196$$

The height of the lighthouse, $DC$, is found by subtracting the height of the cliff from $BD$.

$$DC = BD - BC$$
$$DC = 181.6196 - 100$$
$$DC = 81.6196$$

The height of the lighthouse is 82 ft.

## 3.9. PARALLELOGRAMS

### PROPERTIES OF PARALLELOGRAMS

A **parallelogram** is a quadrilateral whose opposite sides are parallel. All parallelograms have the following properties:

1. Opposite sides are congruent.
2. Opposite angles are congruent.
3. Adjacent angles are supplementary.
4. The diagonals bisect each other.
5. The two diagonals each divide the parallelogram into two congruent triangles.

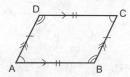  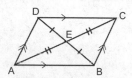

Congruent Sides:
$AB = CD, BC = AD$

Diagonals Bisect Each Other:
$AE = EC, BE = ED$

Congruent angles:
$\angle A \cong \angle C, \angle B \cong \angle D$

Diagonals Form Congruent Triangles:
$\triangle BAD \cong \triangle DCB, \triangle ADC \cong \triangle CBA$

Supplementary angles:
$\angle A$ and $\angle B$, $\angle B$ and $\angle C$
$\angle C$ and $\angle D$, $\angle D$ and $\angle A$

### SPECIAL PARALLELOGRAMS

**Rectangle**—a parallelogram with right angles
**Rhombus**—a parallelogram with consecutive congruent sides
**Square**—a parallelogram with right angles and consecutive congruent sides

Rectangles, rhombuses, and squares have all the properties of parallelograms plus the following properties:

|  | Rectangle | Rhombus | Square |
|---|:---:|:---:|:---:|
| 4 right angles | ✔ |  | ✔ |
| Congruent diagonals | ✔ |  | ✔ |
| 4 congruent sides |  | ✔ | ✔ |
| Perpendicular diagonals |  | ✔ | ✔ |
| Diagonals bisect the angles |  | ✔ | ✔ |

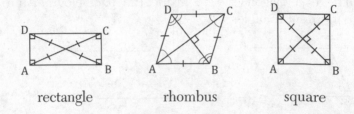

rectangle          rhombus          square

## TRAPEZOIDS

A *trapezoid* is a quadrilateral with exactly one pair of parallel sides. A trapezoid with one pair of parallel sides is shown in the accompanying figure. The parallel sides are called the bases and the nonparallel sides are called the legs. The same side interior angles formed with the parallel bases are supplementary.

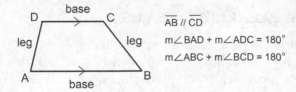

$\overline{AB} \parallel \overline{CD}$
$m\angle BAD + m\angle ADC = 180°$
$m\angle ABC + m\angle BCD = 180°$

## CLASSIFYING PARALLELOGRAMS AND TRAPEZOIDS

The following table lists the properties sufficient to prove a quadrilateral is a parallelogram, rectangle, rhombus, square, or trapezoid.

| Classification | What's Needed to Classify the Figure |
|---|---|
| Trapezoid | One pair of parallel sides |
| Parallelogram | Any *one* of the following:<br><br>• Two pairs of opposite sides parallel<br>• Two pairs of opposite sides congruent<br>• Two pairs of opposite angles congruent<br>• Consecutive angles supplementary<br>• Diagonals that bisect each other<br>• One pair of sides congruent *and* parallel |
| Rectangle | Any one property from the parallelogram list, plus any one of the following:<br>• One right angle<br>• Diagonals that are congruent |
| Rhombus | Any one property from the parallelogram list, plus any one of the following:<br>• Diagonals are perpendicular<br>• One pair of consecutive sides is congruent<br>• A diagonal bisects one of the angles |
| Square | Any one property from the parallelogram list<br>*plus*<br>Any one property from the rectangle list<br>*plus*<br>Any one property from the rhombus list |

The parallelogram properties needed for a proof are illustrated in the following sketches.

| Property of Parallelogram | What It Looks Like |
|---|---|
| Two pairs of opposite sides are parallel. | 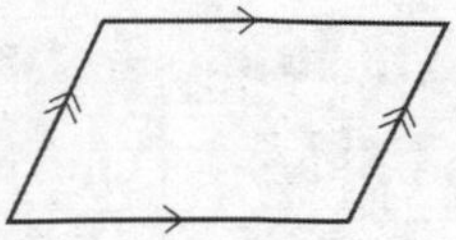 |
| Two pairs of opposite sides are congruent. | 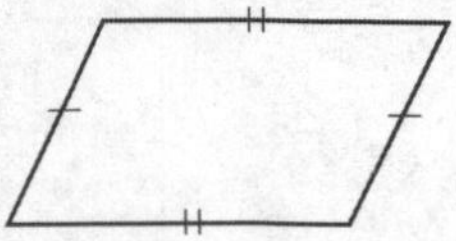 |
| One pair of parallel sides is congruent and parallel. | 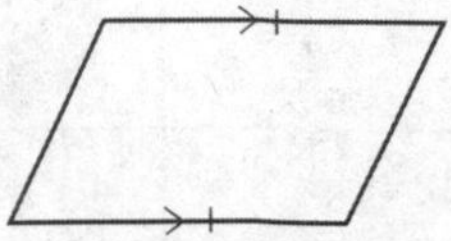 |
| Two pairs of opposite angles are congruent. |  |
| Two pairs of consecutive angles are supplementary. | 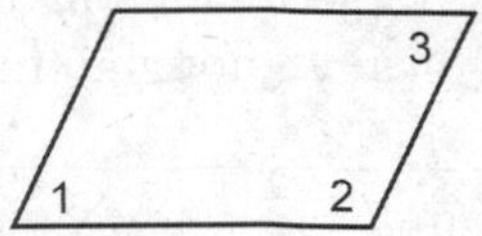 <br>∠1 and ∠2, ∠2 and ∠3<br>are supplementary |
| The diagonals bisect each other. | 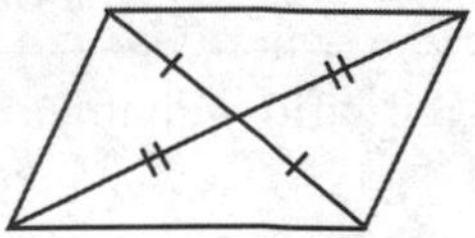 |

# Practice Exercises

1. Parallelogram *PARK* has congruent and perpendicular diagonals. Which of the following describes *PARK*?
   (1) square
   (2) rectangle, but not rhombus
   (3) rhombus, but not rectangle
   (4) trapezoid, but not rectangle or rhombus

2. Quadrilateral *JUMP* has the following properties: $JU = MP$, $UM = PJ$, and $JU = UM$. How can *JUMP* be classified?
   (1) parallelogram only
   (2) parallelogram and rectangle
   (3) parallelogram and rhombus
   (4) not enough information to make any classification

3. In parallelogram *RSTU*, $m\angle R = 8x + 12$ and $m\angle T = 4x + 24$. What is $m\angle R$?
   (1) 144°        (3) 36°
   (2) 72°         (4) 18°

4. In parallelogram *ABCD*, the measures of $\angle B$ and $\angle C$ are in a 3:2 ratio. What is the measure of $\angle B$?
   (1) 120°        (3) 72°
   (2) 108°        (4) 60°

**5.** What are the measures of ∠1 and ∠2 in the given trapezoid?

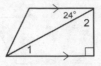

(1) m∠1 = 24° and m∠2 = 24°
(2) m∠1 = 66° and m∠2 = 24°
(3) m∠1 = 66° and m∠2 = 66°
(4) m∠1 = 24° and m∠2 = 66°

**6.** Given rhombus *ABCD* with diagonals $\overline{AC}$ and $\overline{BD}$ intersecting at *E*. If m∠*DAE* = 41°, what is the measure of ∠*CBE*?

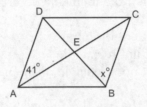

(1) 20.5°          (3) 45°
(2) 41°            (4) 49°

**7.** *ABCD* is a rhombus with diagonals $\overline{AC}$ and $\overline{BD}$ intersecting at *E*. If *AC* = 16 and *BD* = 12, what is the length of $\overline{AB}$?

(1) 14            (3) 10
(2) $4\sqrt{14}$   (4) 28

**8.** Find the measure of ∠C in the parallelogram ABCD.

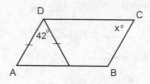

(1) 42°                    (3) 69°
(2) 58°                    (4) 71°

**9.** Diagonal $\overline{QS}$ is drawn in rectangle QRST. If m∠RQS = $(4x + 2)°$ and m∠QSR = $(3x + 11)°$, what is m∠RQS?
(1) 11°                    (3) 46°
(2) 44°                    (4) 90°

**10.** In rectangle CARS, diagonals $\overline{SA}$ and $\overline{CR}$ intersect at X. If m∠RSX = 28°, what is the measure of ∠SXC?

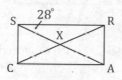

(1) 56°                    (3) 66°
(2) 62°                    (4) 68°

**11.** Square STAR has diagonals that intersect at P. If the perimeter of STAR is 24, what is the length of RP in simplest radical form?

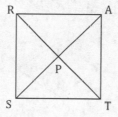

**12.** Given: $\angle E \cong \angle G$ and $\angle EDF \cong \angle GFD$
Prove: *DEFG* is a parallelogram

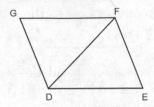

**13.** Given: $\overline{DKB}, \overline{DLC}$, *PLDK* is a rhombus, and $\overline{BK} \cong \overline{CL}$
Prove: $\angle B \cong \angle C$

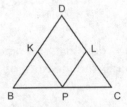

**14.** Given: Rectangle *ABCD* with diagonals $\overline{AC}$ and $\overline{BD}$
Prove: The diagonals of a rectangle are congruent ($\overline{AC} \cong \overline{BD}$)

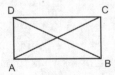

**15.** Given: $\angle BCA \cong \angle DAC$, $\angle BAC \cong \angle BCA$, $\overline{BC} \cong \overline{AD}$
Prove: *ABCD* is a rhombus

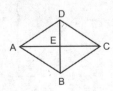

**Solutions**

**1.** A parallelogram with perpendicular diagonals is a rhombus, and a parallelogram with congruent diagonals is a rectangle. A figure with the properties of a rectangle and a rhombus is a square.

The correct choice is **(1)**.

**2.** Opposites sides $\overline{JU}$ and $\overline{MP}$, and $\overline{UM}$ and $\overline{PJ}$ are congruent, making $JUMP$ a parallelogram. Consecutive sides $\overline{JU}$ and $\overline{UM}$ are congruent, making $JUMP$ a rhombus.

The correct choice is **(3)**.

**3.** $\mathrm{m}\angle R = \mathrm{m}\angle T$      Opposite angles of a parallelogram are congruent

$$8x + 12 = 4x + 24$$
$$4x + 12 = 24$$
$$4x = 12$$
$$x = 3$$
$$\mathrm{m}\angle R = 8(3) + 12 \quad \text{Substitute } x = 3$$
$$= 24 + 12$$
$$= 36°$$

The correct choice is **(3)**.

**4.** Let $\mathrm{m}\angle B = 3x$ and $\mathrm{m}\angle C = 2x$, and apply the parallelogram property that consecutive angles are supplementary.

$$\mathrm{m}\angle B + \mathrm{m}\angle C = 180°$$
$$3x + 2x = 180$$
$$5x = 180$$
$$x = 36$$
$$\mathrm{m}\angle B = 3(36) \quad \text{Substitute } x = 36$$
$$= 108°$$

The correct choice is **(2)**.

**5.** The diagonal is a transversal forming a pair of congruent alternate interior angles, so $m\angle 1 = 24°$. The triangle angle sum theorem can be used on the bottom triangle to find $\angle 2$.

$$m\angle 1 + m\angle 2 + 90° = 180°$$
$$24° + m\angle 2 + 90° = 180°$$
$$m\angle 2 + 114° = 180°$$
$$m\angle 2 = 66°$$

The correct choice is (**4**).

**6.** $m\angle BCE = 41°$    $\angle DAE$ and $\angle BCE$ are congruent alternate interior angles

$m\angle BEC = 90°$    Diagonals of a rhombus are perpendicular

$m\angle CBE + m\angle BEC + m\angle BCE = 180°$    Triangle angle sum theorem

$$m\angle CBE + 90° + 41° = 180°$$
$$m\angle CBE + 131° = 180°$$
$$m\angle CBE = 49°$$

The correct choice is (**4**).

**7.** Use the rhombus property that the diagonals bisect each other and are perpendicular.

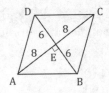

$$AE = \frac{1}{2}AC \qquad \text{Diagonals of a rhombus bisect}$$
$$\text{each other}$$
$$= \frac{1}{2}16$$
$$= 8$$

$$BE = \frac{1}{2}BD$$

$$= \frac{1}{2}(12)$$

$$= 6$$

$\triangle AEB$ is a right triangle     Diagonals of a rhombus are
perpendicular

$(AE)^2 + (BE)^2 = (AB)^2$     Pythagorean theorem

$$6^2 + 8^2 = (AB)^2$$
$$100 = (AB)^2$$
$$AB = 10$$

The correct choice is **(3)**.

**8.** $m\angle A = \dfrac{180° - 42°}{2}$     Isosceles triangle theorem

$$= 69°$$

$m\angle C = m\angle A$     Opposite angles are congruent

$$= 69°$$

The correct choice is **(3)**.

**9.** Start by sketching the figure.

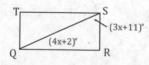

$\triangle QSR$ is a right triangle with a right angle at $\angle R$, so $\angle RQS$ and $\angle QSR$ must be complementary.

$$m\angle RQS + m\angle QSR = 90°$$
$$4x + 2 + 3x + 11 = 90$$
$$7x + 13 = 90$$
$$7x = 77$$
$$x = 11$$

m∠$RQS$ = 4(11) + 2    Substitute $x = 11$
m∠$RQS$ = 46°

The correct choice is **(3)**.

**10.** m∠$CSX$ + m∠$RSX$ = 90°  Angles in a rectangle are right
angles

m∠$CSX$ + 28° = 90°

m∠$CSX$ = 62°

The diagonals of a rectangle are congruent and bisect each other; therefore, $\overline{SX} \cong \overline{CX}$, making $\triangle SXC$ isosceles.

m∠$SCX$ = m∠$CSX$    Base angles of an isosceles triangle
are congruent

= 62°

m∠$CSX$ + m∠$SCX$ + m∠$SXC$ = 180°    Triangle angle sum
theorem

62° + 62° + m∠$SXC$ = 180°
124° + m∠$SXC$ = 180°
m∠$SXC$ = 56°

The correct choice is **(1)**.

**11.** $RS = \dfrac{1}{4}$ perimeter    All sides of a square are congruent

$= \dfrac{1}{4}(24)$

$= 6$

$RP = PS$    Diagonals of a square bisect each
other and are congruent

$(RP)^2 + (PS)^2 = (RS)^2$    Pythagorean theorem
$(RP)^2 + (RP)^2 = 6^2$
$2(RP)^2 = 36$
$(RP)^2 = 18$
$RP = \sqrt{18}$
$= 3\sqrt{2}$

**12.**

| Statement | Reason |
|---|---|
| 1. $\angle EDF \cong \angle GFD$ | 1. Given |
| 2. $\overline{ED} \parallel \overline{FG}$ | 2. Two lines are parallel if the alternate interior angles formed are congruent |
| 3. $\angle E \cong \angle G$ | 3. Given |
| 4. $\overline{FD} \cong \overline{FD}$ | 4. Reflexive property |
| 5. $\triangle EDF \cong \triangle GFD$ | 5. AAS |
| 6. $\angle EFD \cong \angle GDF$ | 6. CPCTC |
| 7. $\overline{GD} \parallel \overline{FE}$ | 7. Two lines are parallel if the alternate interior angles formed are congruent |
| 8. $DEFG$ is a parallelogram | 8. A quadrilateral with two pairs of opposite parallel sides is a parallelogram |

**13.**

| Statement | Reason |
|---|---|
| 1. $\overline{DKB}$, $\overline{DLC}$, and $PLDK$ is a rhombus | 1. Given |
| 2. $\overline{KD} \cong \overline{LD}$ | 2. Consecutive sides of a rhombus are congruent |
| 3. $\overline{BK} \cong \overline{CL}$ | 3. Given |
| 4. $\overline{BK} + \overline{KD} \cong \overline{CL} + \overline{LD}$ | 4. Addition postulate |
| 5. $\overline{BD} \cong \overline{CD}$ | 5. Partition property |
| 6. $\angle B \cong \angle C$ | 6. Angles opposite congruent sides in triangles are congruent |

**14.** The strategy is to prove $\triangle ABC \cong \triangle BAD$ and then apply CPCTC.

| Statement | Reason |
| --- | --- |
| 1. Rectangle $ABCD$ with diagonals $\overline{AC}$ and $\overline{BD}$ | 1. Given |
| 2. $\angle ABC$ and $\angle BAD$ are right angles | 2. All angles in a rectangle are right angles |
| 3. $\angle ABC \cong \angle BAD$ | 3. Right angles are congruent |
| 4. $\overline{AD} \cong \overline{BC}$ | 4. Opposite sides of a rectangle are congruent |
| 5. $\overline{AB} \cong \overline{AB}$ | 5. Reflexive property |
| 6. $\triangle ABC \cong \triangle BAD$ | 6. SAS |
| 7. $\overline{AC} \cong \overline{BD}$ | 7. CPCTC |

**15.** The strategy is to first prove $ABCD$ is a parallelogram and then prove $\triangle ABC$ is an isosceles triangle, with legs $\overline{AB} \cong \overline{BC}$.

| Statement | Reason |
|---|---|
| 1. $\angle BCA \cong \angle DAC$ | 1. Given |
| 2. $\overline{BC} \parallel \overline{AD}$ | 2. Two lines are parallel if the alternate interior angles formed by a transversal are congruent |
| 3. $\overline{BC} \cong \overline{AD}$ | 3. Given |
| 4. $ABCD$ is a parallelogram | 4. A quadrilateral is a parallelogram if one pair of sides is parallel and congruent |
| 5. $\angle BAC \cong \angle BCA$ | 5. Given |
| 6. $\triangle ABC$ is isosceles | 6. A triangle with congruent base angles is isosceles |
| 7. $\overline{AB} \cong \overline{BC}$ | 7. Sides opposite congruent angles in a triangle are congruent |
| 8. $ABCD$ is a rhombus | 8. A parallelogram with a pair of consecutive congruent sides is a rhombus |

## 3.10 COORDINATE GEOMETRY PROOFS

Distance, midpoint, and slope are the tools used on the coordinate plane to prove properties of a figure. They are used as follows:

| Quantity | Formula | Use |
|----------|---------|-----|
| Slope | $m = \dfrac{y_2 - y_1}{x_2 - x_1}$ or $\dfrac{\text{rise}}{\text{run}}$ | Prove segments or lines are parallel (slopes are equal) or perpendicular (slopes are negative reciprocals) |
| Midpoint | $x_{MP} = \dfrac{1}{2}(x_1 + x_2)$ $y_{MP} = \dfrac{1}{2}(y_1 + y_2)$ | Prove segments bisect each other (midpoints are concurrent) |
| Length | $d = \sqrt{(x_2 - x_1)^2 + (y_2 - y_1)^2}$ | Prove segments are congruent (distances between endpoints are equal) |

The steps to writing a good coordinate geometry proof are

- *Graph the points*—The graph will help you plan a strategy, check your work, and help with some calculations.
- *Plan a strategy*—Determine what property you will demonstrate and what calculations are needed.
- *Perform the calculations*—Write down the general equations you are using and clearly label which segments correspond to which calculations.
- *Write a summary statement*—You must state in words a justification that explains why your calculations justify the proof.

# PROVING PARALLELOGRAMS ON THE COORDINATE PLANE

A given quadrilateral can be proven to be a parallelogram by demonstrating one of the parallelogram properties. The three properties that can be easily proven are parallel sides, congruent sides, and diagonals bisecting each other. The calculations required and suggested summary statements are shown in the following table.

| Property | Calculations | Summary Statement |
|---|---|---|
| 2 pairs of opposite sides are parallel | Slope of each side (4 slope calculations) | The slopes of opposite sides are equal; therefore, both pairs of opposite sides are parallel. A quadrilateral with two pairs of opposite sides parallel is a parallelogram. |
| 2 pairs of opposite sides are congruent | Length of each side (4 distance calculations) | The lengths of opposite sides are equal; therefore, both pairs of opposite sides are congruent. A quadrilateral with two pairs of opposite congruent sides is a parallelogram. |
| The diagonals bisect each other | Midpoints of the diagonals (2 midpoint calculations) | The diagonals have the same midpoint; therefore, they bisect each other. A quadrilateral whose diagonals bisect each other is a parallelogram. |

## PROVING SPECIAL PARALLELOGRAMS ON THE COORDINATE PLANE

To prove a quadrilateral is a special parallelogram, a good strategy is to first prove the figure is a parallelogram and then to demonstrate one of the special properties shown in the table. For squares, you need to show that the figure is both a rectangle and a rhombus.

First show the figure is a parallelogram, then ...

| Figure | Property | Calculation | Example of Summary Statement |
|---|---|---|---|
| Rhombus | Perpendicular diagonals | Slope of each diagonal | The slope of the diagonals are negative reciprocals; therefore, they are ⊥. A parallelogram with ⊥ diagonals is a rhombus. |
| | Two consecutive congruent sides | Length of two consecutive sides | The lengths of 2 consecutive sides are equal, so they are ≅. A parallelogram with consecutive ≅ sides is a rhombus. |
| Rectangle | A right angle | Slope of two consecutive sides | The slopes of two consecutive sides are negative reciprocals, so the sides are ⊥ and form a right angle. A parallelogram with a right angle is a rectangle. |
| | Congruent diagonals | Length of each diagonal | The lengths of the diagonals are ≅, so they are congruent. A parallelogram with congruent diagonals is a rectangle. |
| Square | Show the figure is both a rectangle and a rhombus. | | A figure that is both a rectangle and a rhombus is a square. |

One strategy to aid in remembering which properties to prove is to work only with the diagonals:

- Midpoints for a parallelogram
- Midpoints and slope for a rhombus
- Midpoints and length for a rectangle

## OTHER COORDINATE GEOMETRY PROOFS

A quadrilateral can be shown to be a trapezoid using slope calculations to show two sides are parallel.

Triangles can be classified using coordinate geometry calculations:

**Isosceles triangle**—distance calculation to show two sides are congruent

**Equilateral**—distance calculation to show three sides are congruent

**Right triangle**—slope calculation to show two sides are perpendicular

Coordinate geometry can be used to prove special segments in triangles. The following table summarizes some of the segments and points of concurrency, and the calculations needed.

| Segment | Calculation Needed | Summary Statement |
|---|---|---|
| Median | Midpoint | The segment has endpoints at a vertex and midpoint, so it is a median. |
| Altitude | Two slopes | The slopes are negative reciprocals, so the segment is ⊥ to the opposite side, making it an altitude. |
| Perpendicular bisector | Slope and midpoint | The slopes are negative reciprocals, making the segment ⊥, and the segment passes through the midpoint, making it a bisector. Therefore, it is a perpendicular bisector. |
| Midsegment | Two midpoints | The segment's endpoints are the midpoints of sides of the triangle; therefore, it is a midsegment. |
| Circumcenter | Three lengths | The distance from the point to the three vertices are equal, so it is equidistant from each vertex. Therefore, the point is the circumcenter. |
| Centroid | Two lengths | The distances from the point to the vertices and the distance from the point to the opposite midpoint are in a 1:2 ratio; therefore, the point is a centroid. |

# Practice Exercises

1. The coordinates of quadrilateral $BGRM$ are $B(0, 2)$, $G(6, 0)$, $R(9, 9)$, and $M(3, 11)$.
   Prove $BGRM$ is a rectangle.

2. $\triangle GYP$ has coordinates $G(1, 1)$ $Y(7, 5)$ and $P(3, 11)$.
   Prove $\triangle GYP$ is an isosceles right triangle.

3. Given parallelogram $FLIP$ with vertices $F(2, 9)$, $L(h, k)$, $I(11, 6)$, and $P(1, 2)$, find the values of $h$ and $k$.

4. Given quadrilateral $CUTE$ with coordinates $C(-5, -4)$, $U(-3, 4)$, $T(5, 6)$, and $E(3, -2)$, prove $CUTE$ is a rhombus.

5. Quadrilateral $PLOW$ has coordinates $P(1, -4)$, $L(-1, 4)$, $O(7, 6)$, and $W(9, -2)$.
   Prove $PLOW$ is a square.

6. Quadrilateral $ABCD$ has coordinates $A(1, 2)$, $B(9, 4)$, $C(5, 8)$, and $D(1, 7)$.
   Prove $ABCD$ is a trapezoid.

7.  Segment $\overline{MG}$ has coordinates $M(0, 2)$ and $G(4, 0)$. Point $T$ is the midpoint of $\overline{MG}$. $M''G''$ is the image of $\overline{MG}$ after a 90° rotation about $T$ followed by a dilation centered about $T$ with a scale factor of 2.

a) State the coordinates of $M''$ and $G''$ and graph $\overline{M''G''}$.

b) Prove $MG''GM''$ is a rhombus.

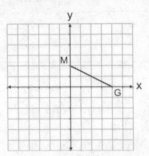

8.  $\triangle ABC$ has coordinates $A(1, 5)$, $B(7, -1)$, and $C(13, 11)$. If point $G$ has coordinates $(7, 5)$, prove $G$ is the centroid of $\triangle ABC$.

9.  $\triangle DOG$ has coordinates $D(4, 3)$, $O(9, 8)$, and $G(1, 12)$. If point $A$ has coordinates $(3, 6)$, prove $\overline{OA}$ is an altitude of $\triangle DOG$.

10. Rectangle $ABCD$ has coordinates $A(4, 12)$, $B(13, 3)$, $C(9, -1)$, and $D(0, 8)$. Points $F$ and $G$ have coordinates $F(2, 10)$ and $G(7, 5)$, and $AHGF$ is also a rectangle. Is rectangle $ABCD$ similar to rectangle $AHGF$? Justify your answer.

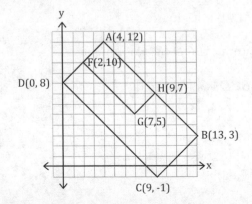

**Solutions**

**1.** There are several different approaches to proving a quadrilateral is a rectangle. Here, we will calculate the slopes of the 4 sides. This will let us show opposite sides have the same slope and are parallel, and consecutive sides have negative reciprocal slopes and are perpendicular. *BGRM* is graphed in the accompanying diagram.

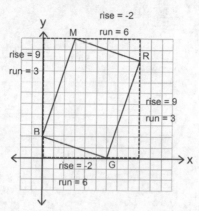

Using the grid, we can calculate slope from slope = $\dfrac{\text{rise}}{\text{run}}$. The rise and run for each side of *BGRM* are shown in the figure.

Slope $\overline{BM}$ = $\dfrac{\text{rise}}{\text{run}}$      Slope $\overline{GR}$ = $\dfrac{\text{rise}}{\text{run}}$

      = $\dfrac{9}{3}$                 = $\dfrac{9}{3}$

      = 3                     = 3

Slope $\overline{BG}$ = $\dfrac{\text{rise}}{\text{run}}$      Slope $\overline{MR}$ = $\dfrac{\text{rise}}{\text{run}}$

      = $\dfrac{-2}{6}$             = $\dfrac{-2}{6}$

      = $-\dfrac{1}{3}$           = $-\dfrac{1}{3}$

The slopes of opposite sides are equal, therefore they are parallel and *BRGM* is a parallelogram.

The slopes of consecutive sides are negative reciprocals, making them perpendicular. $BGRM$ is a rectangle because it is a parallelogram with right angles.

**2.** The strategy is to show that $\overline{GY} \cong \overline{YP}$ using distance calculations and $\overline{GY} \perp \overline{YP}$ using slope calculations.

$$\text{slope} = \frac{y_2 - y_1}{x_2 - x_1}$$

$$\text{slope } \overline{GY} = \frac{5-1}{7-1} \qquad\qquad \text{slope } \overline{YP} = \frac{11-5}{3-7}$$

$$= \frac{4}{6} = \frac{2}{3} \qquad\qquad\qquad = -\frac{6}{4} = -\frac{3}{2}$$

$$\text{length} = \sqrt{(x_1 - x_2)^2 + (y_1 - y_2)^2}$$

$$\text{length } \overline{GY} = \sqrt{(7-1)^2 + (5-1)^2}$$

$$= \sqrt{36 + 16}$$

$$= \sqrt{52}$$

$$\text{length } \overline{YP} = \sqrt{(7-3)^2 + (5-11)^2}$$

$$= \sqrt{16 + 36}$$

$$= \sqrt{52}$$

$\overline{GY}$ is perpendicular to $\overline{YP}$ because their slopes are negative reciprocals, and $\overline{GY} \cong \overline{YP}$ because the lengths $GY$ and $YP$ are equal. Therefore, $\triangle GYP$ is a right isosceles triangle.

**3.** The opposite sides of a parallelogram are parallel and are congruent. The strategy is to locate point $L$ so that $FL = PI$ and the slopes of $\overline{FL}$ and $\overline{PI}$ are equal.

$$\text{slope} = \frac{y_2 - y_1}{x_2 - x_1}$$

$$\text{slope } \overline{PI} = \frac{6-2}{11-1}$$

$$= \frac{4}{10}$$

$$\text{slope } \overline{FL} = \frac{4}{10} \qquad \text{Parallel lines have equal slopes.}$$

Since slope is also equal to $\dfrac{\text{rise}}{\text{run}}$, locate point $L$ by adding the rise to the $y$-coordinate of $F$, and add the run to the $x$-coordinate of $x$.

$$x\text{-coordinate of } L = 2 + 10$$
$$= 12$$

$$y\text{-coordinate of } L = 9 + 4$$
$$= 13$$

The coordinates of $L$ are (12, 13).

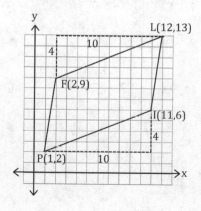

**4.** One strategy is to first prove the figure is a parallelogram by showing the diagonals bisect each other using midpoint calculations. Then prove it is a rhombus by showing the diagonals are perpendicular using slope calculations.

$$\text{midpoint} = \frac{x_1 + x_2}{2}, \frac{y_1 + y_2}{2}$$

midpoint $\overline{CT}$: $\frac{-5+5}{2}, \frac{-4+6}{2}$     midpoint $\overline{EU}$: $\frac{3+(-3)}{2}, \frac{-2+4}{2}$

$(0, 1)$                                              $(0, 1)$

$$\text{slope} = \frac{y_2 - y_1}{x_2 - x_1}$$

slope $\overline{CT} = \frac{6-(-4)}{5-(-5)}$          slope $\overline{EU} = \frac{4-(-2)}{-3-3}$

$\qquad = \frac{10}{10}$                                $\qquad = -\frac{6}{6}$

$\qquad = 1$                                            $\qquad = -1$

The diagonals of *CUTE* bisect each because they have the same midpoint, making it a parallelogram. The diagonals are also perpendicular because their slopes are negative reciprocals. Therefore, *CUTE* is a rhombus.

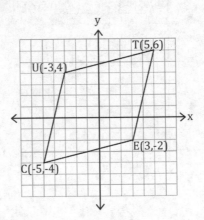

**5.** First prove the figure is a parallelogram by showing the diagonals bisect each other using midpoint calculations; then show it is a rhombus with perpendicular diagonals using slope. Finally, show it is a rectangle with a pair of perpendicular sides using slope.

$$\text{midpoint} = \frac{x_1 + x_2}{2}, \frac{y_1 + y_2}{2}$$

midpoint $\overline{PO}$: $\frac{1+7}{2}, \frac{-4+6}{2}$      midpoint $\overline{WL}$: $\frac{9+(-1)}{2}, \frac{-2+4}{2}$

$(4, 1)$                                             $(4, 1)$

$$\text{slope} = \frac{y_2 - y_1}{x_2 - x_1}$$

slope $\overline{PO} = \frac{6-(-4)}{7-1}$          slope $\overline{WL} = \frac{4-(-2)}{-1-9}$

$\qquad = \frac{10}{6}$                              $\qquad = -\frac{6}{10}$

slope $\overline{PW} = \frac{-2-(-4)}{9-1}$         slope $\overline{LP} = \frac{-4-4}{1-(-1)}$

$\qquad = \frac{2}{8}$                               $\qquad = -\frac{8}{2}$

The diagonals of *PLOW* bisect each other because they have the same midpoint, making it a parallelogram. The diagonals of *PLOW* are perpendicular because their slopes are negative reciprocals, making it a rhombus. *PLOW* has a pair of perpendicular consecutive sides because $\overline{PW}$ and $\overline{LP}$ have slopes that are negative reciprocals, making it a rectangle. *PLOW* is both a rectangle and a rhombus, so it is also a square.

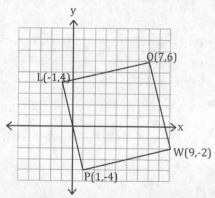

**6.** A good time-saving strategy is to graph the figure first to determine the pair of sides that are likely to be parallel. The parallel sides appear to be $\overline{AB}$ and $\overline{CD}$, and the legs are $\overline{AD}$ and $\overline{BC}$.

$$\text{slope} = \frac{y_2 - y_1}{x_2 - x_1}$$

$$\text{slope } \overline{AB} = \frac{4-2}{9-1} \qquad \text{slope } \overline{CD} = \frac{8-7}{5-1}$$

$$= \frac{2}{8} \qquad\qquad\qquad = \frac{1}{4}$$

$$= \frac{1}{4}$$

$\overline{AB}$ is parallel to $\overline{CD}$ because their slopes are equal, making $ABCD$ a trapezoid.

**7.** a) First find the coordinates of the midpoint $T$ using the midpoint formula; then use $T$ to graphically find the coordinates of $M'$ and $G'$.

$$x_{MP} = \frac{1}{2}(x_1 + x_2)$$

$$= \frac{1}{2}(0 + 4)$$

$$= 2$$

$$y_{MP} = \frac{1}{2}(y_1 + y_2)$$

$$= \frac{1}{2}(2 + 0)$$

$$= 1$$

The coordinates of $T$ are $(2, 1)$.

Rotating a segment 90° about its midpoint results in a segment congruent and perpendicular to the pre-image.

$$\text{slope of } \overline{MG} = \frac{y_2 - y_1}{x_2 - x_1}$$

$$= \frac{0-2}{4-0}$$

$$= -\frac{1}{2}$$

Let $\overline{M'G'}$ be the image of $\overline{MG}$ after the 90° rotation. The slope of $\overline{M'G'}$ is the negative reciprocal, or 2. The coordinates of $M'$ and $G'$ can be found graphically by starting at $T$ and moving with a slope of 2 in either direction for a length equal to $TM$ and $TG$. From the graph, the coordinates are $M'(1, -1)$ and $G'(3, 3)$.

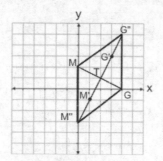

A dilation through midpoint $T$ will extend the segment the same distance on either side of the midpoint because $T$ remains the midpoint. Graph $\overline{M''G''}$ collinear with $\overline{M'G'}$, but twice as long.

b) $MG''GM''$ must be a rhombus. $T$ is the midpoint of both of its diagonals. The diagonals bisect each other making it a parallelogram. The 90° rotation resulted in $\overline{MG} \perp \overline{M''G''}$. A parallelogram with perpendicular diagonals is a rhombus.

**8.** The centroid is the point of concurrency of the medians of a triangle, so one strategy is to graph two medians and look for the point of intersection. A median is the segment from a vertex to the opposite midpoint, so the first step is to find the midpoints of two sides.

$$\text{midpoint} = \frac{x_1 + x_2}{2}, \frac{y_1 + y_2}{2}$$

Let $D$ be the midpoint of $\overline{AB}$: $\dfrac{1+7}{2}, \dfrac{5+(-1)}{2}$

$D(4, 2)$

Let $E$ be the midpoint of $\overline{AC}$: $\dfrac{1+13}{2}, \dfrac{5+11}{2}$

$E(7, 8)$

Graphing $\overline{CD}$ and $\overline{BE}$, the point of intersection is $G(7, 5)$. Therefore, $G$ is the centroid of $\triangle ABC$.

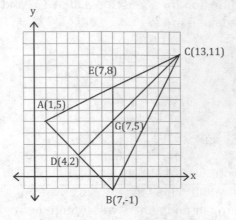

Algebraically, you could have also used the distance formula to calculate $DG$ and $GC$ and then show $DG = \dfrac{1}{2} GC$.

**9.** An altitude is a segment from a vertex perpendicular to the opposite side of a triangle. $\overline{AO}$ is an altitude if it is perpendicular to $\overline{GD}$.

$$\text{slope} = \frac{y_2 - y_1}{x_2 - x_1}$$

$$\text{slope } \overline{AO} = \frac{8-6}{9-3} \qquad\qquad \text{slope } \overline{GD} = \frac{12-3}{1-4}$$

$$= \frac{2}{6} \qquad\qquad\qquad\qquad = -\frac{9}{3}$$

$$= \frac{1}{3} \qquad\qquad\qquad\qquad = -\frac{3}{1}$$

$\overline{AO}$ is perpendicular to $\overline{GD}$ because their slopes are negative reciprocals, so $\overline{AO}$ is an altitude of $\triangle DOG$.

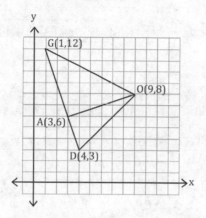

**10.** Since both figures are rectangles, all pairs of corresponding angles must be congruent. The lengths of two pairs of corresponding sides must be calculated to see if they are in the same ratio. The other two pairs must be in the same ratio since opposite sides of a rectangle are congruent.

The corresponding lengths to be checked are $AB$ and $AH$, and $AD$ and $AF$.

$$\text{length} = \sqrt{(x_1 - x_2)^2 + (y_1 - y_2)^2}$$

$$\text{length } AB = \sqrt{(4-13)^2 + (12-3)^2}$$

$$= \sqrt{81+81}$$

$$= \sqrt{162}$$

$$\text{length } AH = \sqrt{(4-9)^2 + (12-7)^2}$$
$$= \sqrt{25+25}$$
$$= \sqrt{50}$$

$$\text{length } AD = \sqrt{(4-0)^2 + (12-8)^2}$$
$$= \sqrt{16+16}$$
$$= \sqrt{32}$$

$$\text{length } AF = \sqrt{(4-2)^2 + (12-10)^2}$$
$$= \sqrt{4+4}$$
$$= \sqrt{8}$$

Calculate the ratios of the two pairs of corresponding sides.

$$\frac{AB}{AH} = \frac{\sqrt{162}}{\sqrt{50}} \qquad\qquad \frac{AD}{AF} = \frac{\sqrt{32}}{\sqrt{8}}$$
$$= \sqrt{\frac{81}{25}} \qquad\qquad\qquad\quad = \sqrt{4}$$
$$= \frac{9}{5} \qquad\qquad\qquad\qquad\quad = 2$$

The rectangles are not similar because the corresponding sides are not in the same ratio.

# 3.11 CIRCLES

## DEFINITIONS AND BASIC THEOREMS
### Segment and Line Definitions in Circles

| Segment or Line | Definition |
|---|---|
| Radius | A segment with one endpoint at the center of the circle and one endpoint on the circle |
| Chord | A segment with both endpoints on the circle |
| Diameter | A chord that passes through the center of the circle |
| Secant | A line that intersects a circle at exactly two points |
| Tangent | A line that intersects a circle at exactly one point |
| Point of tangency | The point at which a tangent intersects a circle |

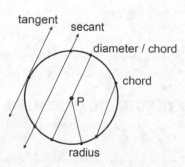

Some basic theorems of circles:

- All radii of a given circle are congruent.
- All circles are similar.
- Two circles are congruent if and only if their radii are congruent.

Since all circles are similar, any circle can be mapped to another using a similarity transformation comprised of a translation to move one center onto the other followed by a dilation to make the radii congruent.

An arc is a section of a circle. An arc may be major, minor, or a semicircle:

- *Minor arc*—an arc spanning less than 180°
- *Semicircle*—an arc spanning exactly 180°
- *Major arc*—an arc spanning more than 180°

When naming a semicircle or major arc, we add an additional point to show which arc is being specified. The accompanying figure shows a minor arc $\overset{\frown}{AB}$, semicircle $\overset{\frown}{ABC}$, and major arc $\overset{\frown}{ABD}$.

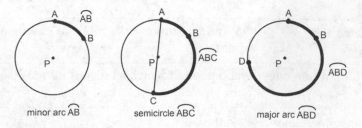

minor arc $\overset{\frown}{AB}$    semicircle $\overset{\frown}{ABC}$    major arc $\overset{\frown}{ABD}$

Some useful arc theorems:

- A diameter intercepts a semicircle.
- The sum of the arc measures around a circle equals 360°.
- The sum of the arc measures around a semicircle equals 180°.

## CENTRAL AND INSCRIBED ANGLES

*Central angle*—an angle whose vertex is the center of a circle and whose rays intersect the circle

---

**Central Angle Theorem**

The angle measure of an arc equals the measure of the central angle that intercepts the arc.

---

*Inscribed angle*—an angle whose vertex is on a circle and whose rays intersect the circle

---

**Inscribed Angle Theorem**
The angle measure of an arc equals twice the measure of the inscribed angle that intercepts the arc.

---

The accompanying figure shows central angle $\angle APB$ and inscribed angle $\angle CMD$. Each angle cuts off, or intercepts, an arc on the circle. Central $\angle APB$ intercepts $\overset{\frown}{AB}$ and inscribed angle $\angle CMD$ intercepts $\overset{\frown}{CD}$.

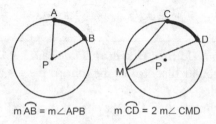

$m \overset{\frown}{AB} = m\angle APB$ $\qquad$ $m \overset{\frown}{CD} = 2\,m\angle CMD$

- Since the inscribed angle measures half the intercepted arc, any inscribed angle that intercepts a semicircle is a right angle.

## CONGRUENT, PARALLEL, AND PERPENDICULAR CHORDS

---

**Congruent Chord Theorem**
Congruent chords intercept congruent arcs on a circle.
$\qquad$ Congruent arcs on a circle are intercepted by congruent chords.

---

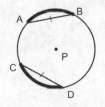

$$\overline{AB} \cong \overline{CD},\ \overset{\frown}{AB} \cong \overset{\frown}{CD}$$

---

*Parallel Chord Theorem*
The two arcs formed *between* a pair of parallel chords are congruent.
    If the two arcs formed *between* a pair of chords are congruent, then the chords are parallel.

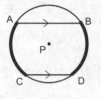

$$\overline{AB} \parallel \overline{CD}, \ \overset{\frown}{AC} \cong \overset{\frown}{BD}$$

---

*Chord–Perpendicular Bisector Theorem*
The perpendicular bisector of any chord passes through the center of the circle.
    A diameter or radius that is perpendicular to a chord bisects the chord.
    A diameter or radius that bisects a chord is perpendicular to the chord.

---

In the accompanying figure, $\overline{AC}$ is perpendicular to diameter $\overline{BD}$ at $E$; therefore, $E$ is the midpoint of $\overline{AC}$.

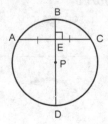

You can think of this theorem linking three properties of a diameter:

- Bisects another chord
- Is perpendicular to another chord
- Is a diameter

If any two are true, then the third must also be true.

---

***Tangent Radius Theorem***
- A diameter or radius to a point of tangency is perpendicular to the tangent.
- A line perpendicular to a tangent at the point of tangency passes through the center of the circle.

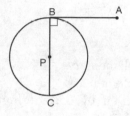

---

***Congruent Tangent Theorem***
Given a circle and external point $Q$, segments between the external point and the two points of tangency are congruent.

---

In the accompanying figure, tangents $\overrightarrow{QA}$ and $\overrightarrow{QB}$ are both constructed from point $Q$, and $\overline{QA} \cong \overline{QB}$.

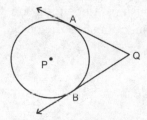

## CYCLIC QUADRILATERALS

Quadrilaterals that are inscribed in a circle are called cyclic quadrilaterals.

---

**Cyclic Quadrilateral Theorem**
Opposite angles of cyclic quadrilaterals are supplementary.

---

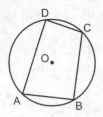

∠A and ∠C are supplementary.
∠B and ∠D are supplementary.

This property is proven by considering the intercepted arcs. Together, arcs $\overset{\frown}{ADC}$ and $\overset{\frown}{ABC}$ comprise the entire circle, so $m\overset{\frown}{ADC} + m\overset{\frown}{ABC} = 360°$. Since the intercepted angles measure $\frac{1}{2}$ the arc measures, $m\angle ABC + m\angle ADC = 180°$, and the angles are supplementary. The same justification can be used to show ∠BAD and ∠BCD are supplementary.

## ANGLES FORMED BY INTERSECTING CHORDS, SECANTS, AND TANGENTS

Angles formed by intersecting chords, secants, and tangents follow three different relationships, depending on whether the vertex of the angle lies within, on, or outside the circle.

| Vertex inside the circle | angle measure = $\frac{1}{2}$ the *sum* of the intercepted arcs |
|---|---|
| 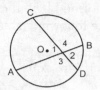 | $m\angle 1 = m\angle 2 = \frac{1}{2}(m\widehat{AC} + m\widehat{DB})$ <br><br> $m\angle 3 = m\angle 4 = \frac{1}{2}(m\widehat{BC} + m\widehat{AD})$ |
| **Vertex on the circle** | angle measure = $\frac{1}{2}$ of the intercepted arc |
| 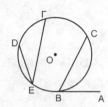 | $m\angle ABC = \frac{1}{2} m\widehat{BC}$ <br><br> $m\angle DEF = \frac{1}{2} m\widehat{DF}$ |
| **Vertex outside the circle** | angle measure = $\frac{1}{2}$ the *difference* of the intercepted arcs |
| 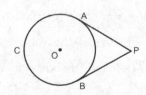 | $m\angle P = \frac{1}{2}(m\widehat{ACB} - m\widehat{AB})$ |
| 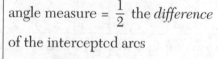 | $m\angle P = \frac{1}{2}(m\widehat{CD} - m\widehat{AB})$ |
| 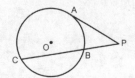 | $m\angle P = \frac{1}{2}(m\widehat{AC} - m\widehat{AB})$ |

# SEGMENT RELATIONSHIPS IN INTERSECTING CHORDS, TANGENTS, AND SECANTS

The relationships between segment lengths formed by intersecting chords, tangents, and secants are shown in the following table.

| **Intersecting chords** | products of the parts are equal $$a \cdot b = c \cdot d$$ |
|---|---|
| | |
| **Intersecting secants** | outside · whole = outside · whole $$PA \cdot PC = PB \cdot PD$$ |
| | |
| **Intersecting tangent and secant** | outside · whole = outside · whole $$PA^2 = PB \cdot PC$$ |
| | |

# RADIAN MEASURE, ARC LENGTH, AND SECTORS

The **radian** is a unit of angle measure. $2\pi$ radians is equal to one complete revolution around a circle, or $360°$. Convert between units using:

$$\text{radians} = \frac{\pi}{180°} \cdot \text{degrees}$$

$$\text{degrees} = \frac{180°}{\pi} \cdot \text{radians}$$

- Besides having an angle measure, an arc has a length. The arc and its central angle also enclose a region called a sector.

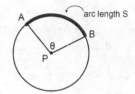

| Arc with length $S$ | Sector $APB$ |
| :---: | :---: |
| intercepted by angle $\theta$ | intercepted by an angle $\theta$ |

Arc length and sector area can be calculated from the measure of the central angle and the radius of the circle using the following formulas:

| Unit of Angle Measure | Arc Length | Sector Area |
| :---: | :---: | :---: |
| **D**egrees | $\dfrac{\pi}{180°} R \cdot \theta$ | $\dfrac{1}{360°} \pi R^2 \cdot \theta$ |
| **R**adians | $R \cdot \theta$ | $\dfrac{1}{2} R^2 \cdot \theta$ |

$\theta$ is the measure of the central angle.

# Practice Exercises

1. Which of the following is a precise definition of a circle?
   (1) the set of points a fixed distance away from a given point
   (2) the set of points equidistant from the two endpoints of a given diameter
   (3) the set of points equidistant from a given center point and a given line
   (4) a closed figure with no angles

2. $\triangle STY$ is inscribed in circle $P$. Which of the following is *not* *necessarily* true?
   (1) Point $P$ is equidistant from $S$, $T$, and $Y$.
   (2) The perpendicular bisector of $\overline{ST}$ passes through $P$.
   (3) The measure of angle $Y$ equals one-half the measure of $\overset{\frown}{ST}$.
   (4) A tangent to the circle at $S$ forms a right angle with $\overline{ST}$.

3. $\triangle FLY$ is inscribed in circle $O$. If $FL = LY$ and $m\overset{\frown}{FY} = 80°$, what is the measure of $\angle F$?
   (1) $10°$          (3) $70°$
   (2) $40°$          (4) $80°$

4. In ⊙*P*, ∠*APB* is a central angle and ∠*ADB* is an inscribed angle. If m∠*APB* = 110°, find m∠*ADB*.

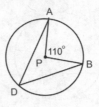

  (1) 55°             (3) 70°
  (2) 60°             (4) 80°

5. The clock in the tower of the Ridgefield town hall is in the shape of a circle 15 feet in diameter. What is the length of the arc around the clock whose central angle is defined by the hour and minute hands at 4:00 P.M.?

  (1) 15.7 ft        (3) 31.4 ft
  (2) 18.6 ft        (4) 60 ft

6. In circle Z, arc $\overset{\frown}{KP}$ is intercepted by central angle ∠*KZP*. If m∠*KZP* = 1.2 radians and *ZK* = 7 cm, what is the length of $\overset{\frown}{KP}$?
  (1) 5.8 cm        (3) 10.1 cm
  (2) 8.4 cm        (4) 26.4 cm

7. Circle *O* has a radius of 14 inches. If ∠*HOT* is a central angle in circle *O* that measures 132°, what is the area of sector *HOT*?

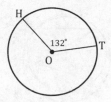

(1) 4.7 in²          (3) 226 in²

(2) 10 in²           (4) 294 in²

8. Given ⊙*P* with radius $\overline{PB}$, tangent $\overline{AB}$, and m∠*P* = 53°, what is the measure of ∠*A*?

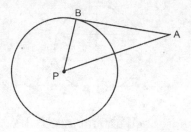

(1) 26.5°          (3) 53°

(2) 37°           (4) 60°

9. In ⊙*P*, chords $\overline{AB}$ and $\overline{CD}$ are parallel. If m$\overset{\frown}{AB}$ = 118° and m$\overset{\frown}{CD}$ = 78°, find m$\overset{\frown}{AC}$.

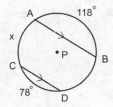

10. In ⊙P, chords $\overline{AB}$ and $\overline{CD}$ are congruent, m$\overarc{BC}$ = 60°, and m$\overarc{CD}$ = 80°. Find m∠ABD.

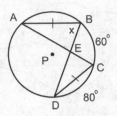

11. $\overline{BDP}$ is a radius of ⊙P and is perpendicular to $\overline{AC}$. If $AC$ = 24 and $PC$ = 13, find $PD$ and $DB$.

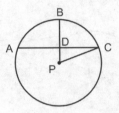

12. △RST is circumscribed about circle P. If SX = 6, XR = 7, and TZ = 8, what is the perimeter of △RST?

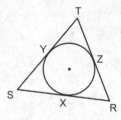

**13.** Secant $\overline{BEA}$ intersects circle $O$ at $E$ and $A$. Secant $\overline{BDR}$ intersects circle $O$ at points $D$ and $R$. If $m\overset{\frown}{AR}$ = 78° and $m\overset{\frown}{ED}$ = 35°, what is the measure of $\angle B$?

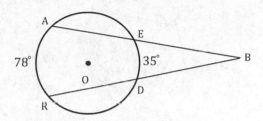

**14.** Secant $\overline{FGH}$ intersects circle $O$ at $G$ and $H$. Secant $\overline{FPQ}$ intersects circle $O$ at $P$ and $Q$. If $FG$ = 6, $GH$ = 10, and $FP$ = 8, find the length of $PQ$.

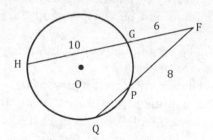

**15.** Tangents $\overline{TR}$ and $\overline{TQ}$ are drawn from point $T$ to circle $A$. $\overline{RAS}$ is a diameter of circle $A$ and $m\angle T$ = 46°, what is the measure of $\overset{\frown}{SQ}$?

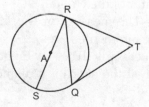

**16.** In $\odot O$, chords $\overline{HLI}$ and $\overline{KLJ}$ intersect at $L$. Prove: $HL \cdot LI = KL \cdot LJ$

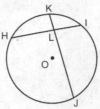

## Solutions

**1.** Examine each choice:

(1) Correct—The fixed distance is the radius of the circle, and the given point is the center of the circle.

(2) Incorrect—The set of points equidistant from the two endpoints of a diameter would be the perpendicular bisector of the diameter.

(3) Incorrect—The set of points equidistant from a center point and line would be a parabola.

(4) Incorrect—Circles are closed figures with no angles, but not all closed figures with no angles are circles. A counterexample would be an ellipse.

The correct choice is (**1**).

**2.** Choice (1) is true—The center of the circumscribed circle is equidistant from the vertices of the triangle.

Choice (2) is true—A perpendicular bisector of a chord always passes through the center of the circle.

Choice (3) is true—The measure of the inscribed angle equals half the measure of the intercepted arc.

Choice (4) is not necessarily true—A tangent will make a right angle with a chord only if the chord is a diameter. $\overline{ST}$ is not necessarily a diameter.

The correct choice is (**4**).

**3.** Sketching the isosceles triangle and labeling the known arc, we see $\angle L$ is the inscribed angle that intercepts the 80° arc.

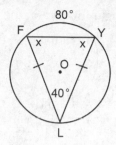

An inscribed angle measures half the intercepted arc:

$$m\angle L = \frac{1}{2}\,m\widehat{FY}$$
$$= \frac{1}{2}\,(80°)$$
$$= 40°$$

Since $FL = LY$, $m\angle F = m\angle Y$ from the isosceles triangle theorem, and we can use the angle sum theorem to find the measures.

$$m\angle F + m\angle L + m\angle Y = 180°$$
$$2x + 40° = 180°$$
$$2x = 140°$$
$$x = 70°$$
$$m\angle F = 70°$$

The correct choice is **(3)**.

**4.** The strategy is to use the central angle to find $m\widehat{AB}$, and then use the arc to find the inscribed angle.

$m\widehat{AB} = m\angle APB = 110°$      The central angle and arc have equal measure.

$m\angle ADB = \frac{1}{2}\,m\widehat{AB} = 55°$      The inscribed angle measures half the intercepted arc.

The correct choice is **(1)**.

**5.** The numbers on the face of the clock divide the circle into 12 congruent central angles. Each central angle measures $\dfrac{360°}{12}$, or 30°. Therefore, the angle formed by the two hands on the clock measures $4 \cdot 30°$, or 120°. The radius $= \dfrac{1}{2}(15)$, or 7.5 ft. Using the formula for arc length,

$$S = 2\pi R \, \frac{\theta}{360°}$$
$$= 2\pi(7.5 \text{ ft})\frac{120°}{360°}$$
$$= 15.7 \text{ ft}$$

The correct choice is (**1**).

**6.** Arc length, $S$, is given by the equation $S = R\theta$, where $\theta$ is the measure of the central angle in radians.

$$S = R\theta$$
$$= 7 \text{ cm} \cdot 1.2$$
$$= 8.4 \text{ cm}$$

The correct choice is (**2**).

**7.** The area of a sector is given by $A = \dfrac{\theta}{360°}\pi R^2$, where $\theta$ is the central angle measured in degrees.

$$A = \frac{132°}{360°}\pi \cdot (14 \text{ in})^2$$
$$= 225.775 \text{ in}^2$$
$$= 226 \text{ in}^2$$

The correct choice is (**3**).

**8.** $\qquad$ m$\angle B = 90°$ $\qquad$ radius perpendicular to a tangent
at the point of tangency

$$m\angle P + m\angle B + m\angle A = 180°$$ $\qquad$ triangle angle sum theorem
$$53° + 90° + m\angle A = 180°$$
$$143° + m\angle A = 180°$$
$$m\angle A = 37°$$

The correct choice is **(2)**.

**9.** Arcs $\overset{\frown}{AC}$ and $\overset{\frown}{DB}$ are the congruent arcs between a pair of parallel chords.

$$m\overset{\frown}{AB} + m\overset{\frown}{BD} + m\overset{\frown}{CD} + m\overset{\frown}{AC} = 360°$$ $\qquad$ circle–arc sum
theorem

$$118° + x + 78° + x = 360°$$ $\qquad$ parallel chord
$$196° + 2x = 360°$$ $\qquad$ theorem
$$2x = 164°$$
$$x = 82°$$
$$m\overset{\frown}{AC} = 82°$$

**10.** $\angle ABD$ is an inscribed angle, so its measure will be half of m$\overset{\frown}{AD}$.

$$m\overset{\frown}{AB} + m\overset{\frown}{BC} + m\overset{\frown}{CD} + m\overset{\frown}{AD} = 360°$$ $\qquad$ circle–arc sum
theorem

$$m\overset{\frown}{AB} = m\overset{\frown}{CD} = 80°$$ $\qquad$ congruent chord
theorem

$$80° + 60° + 80° + m\overset{\frown}{AD} = 360°$$
$$220° + m\overset{\frown}{AD} = 360°$$
$$m\overset{\frown}{AD} = 140°$$
$$m\angle ABD = \frac{1}{2} m\overset{\frown}{AD}$$
$$= 70°$$

**11.** Radius $\overline{BDP}$ is perpendicular to $\overline{AC}$, so it must also bisect $\overline{AC}$. The strategy is to calculate $DC$ and then apply the Pythagorean theorem in $\triangle PDC$.

$$DC = \frac{1}{2} AC$$

$$DC = \frac{1}{2}(24)$$

$$DC = 12$$

| | |
|---|---|
| $(PD)^2 + (DC)^2 = PC^2$ | Pythagorean theorem |
| $(PD)^2 + 12^2 = 13^2$ | |
| $(PD)^2 + 144 = 169$ | |
| $(PD)^2 = 25$ | |
| $PD = 5$ | |

| | |
|---|---|
| $PB = PC = 13$ | all radii of a circle are congruent |
| $PD + DB = PB$ | partition |
| $5 + DB = 13$ | |
| $DB = 8$ | |

**12.** $SY = SX$, $TY = TZ$, and $ZR = XR$ because each pair are tangents from the same point.

The perimeter is, therefore, $2 \cdot 6 + 2 \cdot 7 + 2 \cdot 8 = 42$.

**13.** The angle formed by two secants is equal to half the difference of the intercepted arcs.

$$m\angle B = \frac{1}{2}(m\widehat{AB} - m\widehat{ED})$$

$$= \frac{1}{2}(78° - 35°)$$

$$= 21.5°$$

**14.** When two tangents from the same point intersect a circle, the segments formed follow the relationship:

$$\text{outside} \cdot \text{whole} = \text{outside} \cdot \text{whole}$$

$$FG \cdot FH = FP \cdot FQ$$
$$6 \cdot (6 + 10) = 8 \cdot (8 + PQ)$$
$$6 \cdot 16 = 8(8 + PQ)$$
$$96 = 64 + 8PQ$$
$$32 = 8PQ$$
$$PQ = 4$$

**15.** This problem requires applying multiple circle concepts. The strategy is to first use the fact $\triangle TRQ$ is isosceles to find m$\angle TRQ$ and m$\angle SRQ$ and then to apply the relationship that the measure of $\overset{\frown}{SQ}$ is twice the inscribed angle that intercepts it.

$\overline{TR} \cong \overline{TQ}$ 　　　　　　　　congruent tangent theorem

$\text{m}\angle TRQ = \dfrac{1}{2}(180° - \text{m}\angle T)$ 　　isosceles triangle theorem in $\triangle TRQ$

$\text{m}\angle TRQ = \dfrac{1}{2}(180° - 46°)$

$= 67°$

$\text{m}\angle SRT = 90°$ 　　　　　tangent $RT \perp$ diameter $\overline{RAS}$

$\text{m}\angle SRQ + \text{m}\angle TRQ = 90°$ 　　$\angle SRQ$ and $\angle TRQ$ are complementary

$\text{m}\angle SRQ + 67° = 90°$

$\text{m}\angle SRQ = 23°$

$\text{m}\overset{\frown}{SQ} = 2\text{m}\angle SRQ$ 　　　inscribed angle theorem

$\text{m}\overset{\frown}{SQ} = 2(23°)$

$\text{m}\overset{\frown}{SQ} = 46°$

**16.** The strategy is to sketch $\overline{HJ}$ and $\overline{KI}$, show $\triangle HLJ \sim \triangle KLI$, and then write a proportion between corresponding sides.

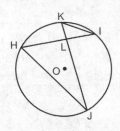

| Statements | Reasons |
|---|---|
| 1. Chords $\overline{HLI}$ and $\overline{KLJ}$ intersect at $L$ | 1. Given |
| 2. Construct $\overline{IK}$ and $\overline{HJ}$ | 2. Two points define a segment |
| 3. $\angle H \cong \angle K$ <br> $\angle J \cong \angle I$ | 3. Inscribed angles that intercept the same arc are congruent |
| 4. $\triangle HLJ \sim \triangle KLI$ | 4. AA |
| 5. $\dfrac{HL}{KL} = \dfrac{LJ}{LI}$ | 5. Corresponding parts of similar triangles are proportional |
| 6. $HL \cdot LI = KL \cdot LJ$ | 6. Cross products of a proportion are equal |

## 3.12 SOLIDS

### DEFINITIONS

- A *solid* is any 3-D figure that is fully enclosed.
- A *face* of a solid is any of the surfaces that bound the solid.
- A *polyhedron* is any solid whose faces are polygons.
- An *edge* is the intersection of two faces in a polyhedron.

### PRISMS

A *prism* is a polyhedron with two congruent, parallel polygons for bases. The bases of a prism can have any shape. The accompanying figure shows 6 different prisms.

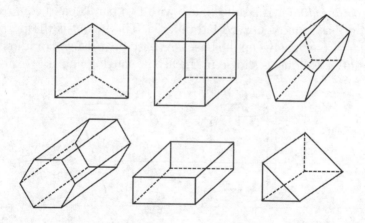

When working with prisms, keep in mind the following facts:

- The height of a prism, $h$, is the distance between the two bases shown in the accompanying figure.
- The lateral faces are all the faces other than the two parallel bases.
- A right prism has lateral edges that are perpendicular to the bases and lateral faces that are rectangles.
- The lateral edges of an oblique prism are not perpendicular to the bases, and the lateral faces are parallelograms.

- The volume of any prism is found with the formula $V = Bh$, where $B$ is the area of the base.

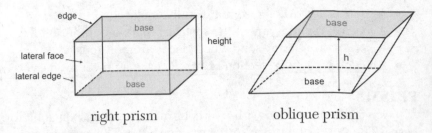

right prism                    oblique prism

## CYLINDERS

A *circular cylinder* is a solid figure with two parallel and congruent circular bases and a curved lateral area. The height is the perpendicular distance between the bases. As with the prisms, cylinders can be right or oblique as shown in the accompanying figure.

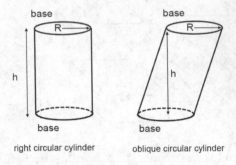

right circular cylinder       oblique circular cylinder

The volume of a cylinder is given by $V = Bh$, where $B$ is the area of the base. In a circular cylinder, the circular base has an area of $\pi R^2$, so the volume formula can be rewritten as $V = \pi R^2 h$.

## CONES AND PYRAMIDS

A *circular cone* is a solid with one circular base that comes to a point at an apex. A *pyramid* is a polyhedron having one polygonal base and triangles for lateral faces. The base of a pyramid can be any polygon, and the lateral faces are all triangles. The height of cones and pyramids is the perpendicular distance from the apex to the base. The slant height is the distance along the lateral surface perpendicular to the perimeter of the base.

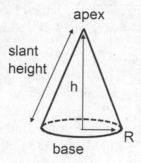

The volume of pyramids and cones are found from the same formula:

$V = \dfrac{1}{3}Bh$ where $B$ is the area of the base.

## SPHERES

A *sphere* is the set of points a fixed distance from a center point, and its volume is given by the formula $V = \dfrac{4}{3}\pi R^3$.

## SURFACE AREA AND LATERAL AREA

The *surface area* is the area of all faces of a solid. Surface area can be found by calculating the area of all the faces of a solid individually and then summing them. You should be able to calculate the surface area of cubes, prisms, and pyramids since the faces of these solids are all polygons. Surface area of cones, cylinders, and spheres involve curved surfaces and are outside the scope of this course.

The lateral area of a solid excludes the bases. For a prism, do not include the two parallel bases. For a pyramid, exclude the one base.

*Lateral face*—any face of a solid other than its bases
*Lateral area*—the area of all the lateral faces of a solid
*Surface area*—the total area of a solid, lateral area + area of bases

## CROSS-SECTIONS

A cross-section of a solid is the two-dimensional figure created when a plane intercepts a solid. Cross-sections are often taken parallel or perpendicular to the base of a figure. The shape of the cross-section depends on the angle at which the plane intersects the solid.

Cross-sections of a pyramid, sphere, and cylinder are shown in the accompanying figure.

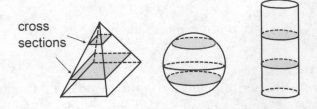

cross sections

A cross-section of a pyramid taken perpendicular to the base will be shaped triangular, as shown in the figure, or trapezoidal. The cross-section of a cylinder perpendicular to the base is a rectangle.

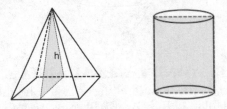

## SOLIDS OF REVOLUTION

Some solids can be generated by rotating a planar figure around a line. These are called *solids of revolution*. Some examples of solids of revolution and the figures that generate them are given in the following table.

| Figure Rotated | Solid |
|---|---|
| Right triangle, <br><br> rotate 360° about leg $\overline{AB}$ | Cone 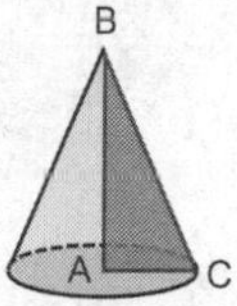 <br><br> |
| Rectangle $ABCD$, <br><br> rotate 360° around side $\overline{CD}$ | Cylinder 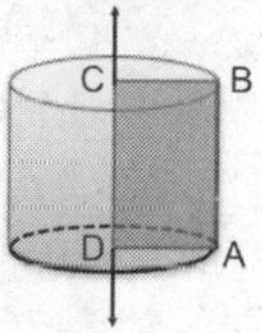 <br><br> |
| Circle $P$, <br><br> rotate 180° about diameter $\overline{AB}$ | Sphere 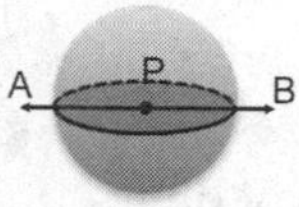 <br><br> |

## CAVALIERI'S PRINCIPLE

If two solids are contained between two parallel planes, and every parallel plane between these two planes intercepts regions of equal area, then the solids have equal volume. Also, any two parallel planes intercept two solids of equal volume.

Both the triangular prism and the rectangular prism shown in the accompanying figure are bounded by planes $R$ and $T$. "Bounded" means that the solids have at least one point contained in each of planes $R$ and $T$, and that the solids do not pass through to the other sides of the planes. As plane $S$ sweeps upward from $R$ toward $T$, corresponding cross-sections are formed in the triangular prism and the rectangular prism. If every pair of corresponding cross-sections are equal in area (though not necessarily congruent), then the two solids have the same volume.

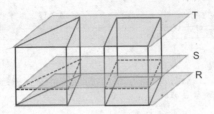

The second part of Cavalieri's principle is illustrated by parallel planes $R$ and $S$. If the conditions of Cavalieri's principle are true, then the two volumes between planes $R$ and $S$ are equal.

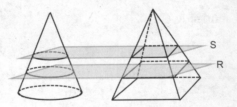

**Example:**

Two stacks of quarters are shown in the accompanying figure. If each stack contains 7 identical quarters, what must be true about the volume of the two stacks? Explain your answer in terms of Cavalieri's principle.

*Solution*:
Each of the quarters has a uniform and congruent cross-section, and any plane parallel to the bases of the two stacks must intercept cross-sections of equal area. Cavalieri's principle states that the volumes of the stacks must be equal.

## JUSTIFICATION OF THE AREA AND VOLUME FORMULAS

### Circumference of a Circle

The formula for the area of a circle can be derived using a limit argument by considering a polygon inscribed in a circle and letting the number of sides increase toward infinity.

Consider a polygon inscribed in circle $O$, as shown in the figure. The central angle, $\angle AOB$, has a measure equal to $\dfrac{360°}{n}$, where $n$ is the number of sides. Altitude $\overline{OP}$ will form $\triangle AOP$, where $m\angle AOP$ equals half the central angle, or $\left(\dfrac{360°}{2n}\right)$, and $AO$ has a length equal to the radius of the circle, $R$. Length $AP$ can be found by applying the sin ratio.

$$\sin(\theta) = \frac{\text{opposite}}{\text{hypotenuse}}$$

$$\sin\left(\frac{360°}{2n}\right) = \frac{AP}{R}$$

$$AP = R \sin\left(\frac{360°}{2n}\right)$$

The perimeter of the polygon is equal to $2n \cdot AP$, since there are $n$ sides and $AP$ is equal to half a side length.

$$\text{Perimeter} = 2n \cdot AP$$

$$= 2n\,R\,\sin\left(\frac{360°}{2n}\right)$$

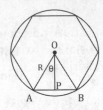

In the limit as the number of sides of the polygon approaches infinity, $n$ approaches infinity, and the perimeter approaches the circumference of the circle. The accompanying table shows the value of $n \sin\left(\dfrac{360°}{2n}\right)$ for increasing values of $n$. As $n$ approaches infinity, $n \sin\left(\dfrac{360°}{2n}\right)$ approaches $\pi$.

| number of sides $n$ | $n \sin\left(\dfrac{360°}{2n}\right)$ |
|:---:|:---:|
| 6 | 3 |
| 20 | 3.1286 |
| 100 | 3.1410 |
| 1000 | 3.1415 |
| Infinite | $\pi$ |

The previous equation for perimeter can be rearranged to

$$\text{Perimeter} = 2n \sin\left(\frac{360°}{2n}\right)R$$

- Substituting circumference for perimeter and $\pi$ for $n \sin\left(\dfrac{360°}{2n}\right)$ gives the familiar result, circumference = $2\pi R$

## Area of a Circle

The area of a circle can also be derived from a limit argument and inscribed polygons. The formula for the area of a polygon is $A = \frac{1}{2}$ perimeter $\times$ apothem. As the number of sides increases toward infinity, the perimeter approaches the circumference of the circle, and the apothem approaches the radius of the circle.

$$A = \frac{1}{2} \text{ perimeter} \times \text{apothem} \rightarrow \frac{1}{2}(2\pi R)R$$

$$A_{\text{circle}} = \frac{1}{2}(2\pi R)R = \pi R^2$$

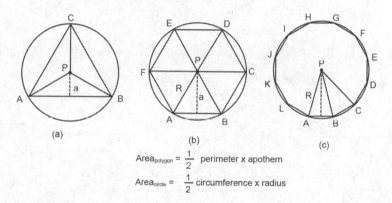

(a)          (b)          (c)

$$\text{Area}_{\text{polygon}} = \frac{1}{2} \text{ perimeter x apothem}$$

$$\text{Area}_{\text{circle}} = \frac{1}{2} \text{ circumference x radius}$$

## Volume of a Cylinder

The formula for volume of a cylinder can be derived from the formula for the volume of a prism. As the number of sides of the base of the prism increases toward infinity, the base takes the shape of a circle with area $\pi R^2$ resulting in the cylinder formula of $\pi R^2 h$.

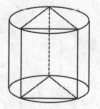
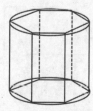

## Volume of a Pyramid

The volume of a square-based pyramid can be derived by *dissection* of a cube with side length $s$, shown in the accompanying figure (a similar method can be used for pyramids of other bases). Let point $P$ be the center of the cube. The four bottom vertices and point $P$ form a square-based pyramid. Five more congruent pyramids can be formed using point $P$ and four vertices at the top, front, back, left, and right. None of the pyramids overlap, so the volume of the cube must be equal to the volume of the 6 pyramids:

$$6\,\text{Volume}_{\text{pyramid}} = \text{Volume}_{\text{cube}} = s^3$$

$$\text{Volume}_{\text{pyramid}} = \frac{1}{6}s^3$$

The height of each pyramid is equal to $\frac{1}{2}s$, or $s = 2h$. Substitute to get

$$\text{Volume}_{\text{pyramid}} = \frac{1}{6}s^3$$

$$= \frac{1}{6}s^2 \cdot s$$

$$= \frac{1}{6}(s^2)2h$$

$$= \frac{1}{3}(s^2)h$$

The area of the base of each pyramid is the same as the area of the base of the cube, or $s^2$, so $s^2$ can be replaced with $B$ to give:

$$\text{Volume}_{\text{pyramid}} = \frac{1}{3}Bh$$

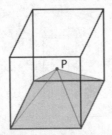 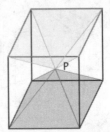

## Volume of a Cone

The same approach that was used for the volume of a cylinder can be applied to derive the formula for the volume of a cone. Starting with a pyramid, whose volume is $\frac{1}{3}Bh$, let the number of sides in the polygonal base approach infinity. The shape of the base will approach a circle, but the height will be unchanged, so the formula remains the same:

$$V_{cone} = \frac{1}{3}Bh$$

## MODELING WITH SOLIDS

Many physical objects can be modeled using the basic solids discussed in this chapter. Some examples are:

*Cylinders*—trees, cans, barrels, people
*Spheres*—balls, balloons, planets
*Prisms*—bricks, boxes, aquariums, swimming pools, rooms, books

Volumes or areas can also be used to calculate some other quantity. Some common relationships to look for are

- Mass = density × volume
- Total cost = cost per unit volume × volume

    or

    Total cost = cost per unit area × area
- Energy contained in a material = heat content × volume
- Total population = population density × area

For example, if a can has a volume of 100 cm³ and is filled with a liquid whose density is 1.2 grams/cm³, then the mass of the liquid inside the can is

$$\text{Mass} = \text{density} \cdot \text{volume}$$
$$\text{Mass} = 1.2 \text{ grams/cm}^3 \cdot 100 \text{ cm}^3$$
$$= 120 \text{ grams}$$

In geometry design problems, we are asked to find the dimensions of some figure or solid that lets the object meet some given condition. Conditions that may be specified include

- maximum or minimum size
- a specified size
- maximum or minimum cost

Approach these problems in the same way as any other modeling problem. The difference comes in using the model to find a maximum or minimum. For example, you may have an equation that gives volume in terms of a length. A maximum or minimum can be found by

- *Trial and error/table*—Using a best first guess for the variable, evaluate the quantity. Change the value of the variable and evaluate the quantity again—it will increase or decrease. Continue until you see the quantity change from decreasing to increasing (a minimum) or increasing to decreasing (a maximum). The table feature of the graphing calculator is excellent for calculating many trials quickly. The table increment can be set to the desired precision.
- *Graphing*—The equation that described the quantity can be graphed, and the maximum/minimum feature of the calculator can then be applied to the graph.
- *Axis of symmetry*—If the equation that describes the quantity is *quadratic* in the form $y = ax^2 + bx + c$, then the maximum or minimum will always occur at the axis of symmetry $x = -b/2a$.

# Practice Exercises

1. A basketball has a diameter of 9.6 inches. What is its volume?
   - (1) 72.3 in$^3$
   - (2) 347 in$^3$
   - (3) 463 in$^3$
   - (4) 3706 in$^3$

2. A right circular cylinder has a volume of 200 cm$^3$. If the height of the cylinder is 6 cm, what is the radius of the cylinder? Round your answer to the nearest hundredth.
   - (1) 2.9 cm
   - (2) 3.0 cm
   - (3) 3.3 cm
   - (4) 3.8 cm

3. Jenna is preparing to plant a new lawn and has a pile of top-soil delivered to her house. A dump truck delivers the topsoil and dumps it in a pile that is cone shaped. The radius of the pile is 9 ft, and the height is 3 ft. If the topsoil has a density of 95 lb/ft$^3$, what is the weight of the topsoil?
   - (1) 18,764 lb
   - (2) 24,175 lb
   - (3) 28,546 lb
   - (4) 32,676 lb

4. A right circular cylinder has a radius of 4 and a height of 10. What is the area of a cross-section taken parallel to, and half-way between, the bases? Write your answer in terms of $\pi$.
   - (1) 4$\pi$
   - (2) 8$\pi$
   - (3) 12$\pi$
   - (4) 16$\pi$

5.  A cylinder and a cone have the same volume. If each solid has the same height, what is the ratio of the radius of the cylinder to the radius of the cone?

    (1) 2:1                         (3) $\sqrt{2}$:2

    (2) 3:1                         (4) $\sqrt{3}$:3

6.  $\triangle JKL$ is a right triangle with a right angle at $K$, $JK = 6$, and $KL = 10$. Which of the following solids is generated with $\triangle JKL$ rotated 360° about $KL$?
    (1) a cylinder with a height of 10 and a diameter of 12
    (2) a cylinder with a height of 6 and a diameter of 20
    (3) a right circular cone with a height of 10 and a diameter of 12
    (4) a right circular cone with a height of 6 and a diameter of 20

7.  Solids $A$ and $B$, shown in the figure below, each have uniform cross-sections perpendicular to the shaded bases.

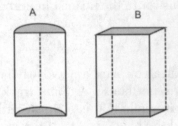

    If the heights of the two solids are equal, which of the following statements represents an application of Cavalieri's principle?
    (1) The surface areas of solid $A$ and solid $B$ may be different even though their volumes are equal.
    (2) Solid $A$ can be generated by rotating a rectangle, wheras solid $B$ is not a solid of revolution.
    (3) If the area of base $A$ equals the area of base $B$, then the volumes of the solids are equal.
    (4) If the volume and height of solid $A$ and $B$ are equal, then their surface areas must be equal.

8. The base of a cone is described by the curve $x^2 + y^2 = 16$. If the altitude of the cone intersects the center of its base, and the volume of the cone is $72\pi$, what is the height of the cone?

9. A cylindrical piece of metal has a radius of 15 in and a height of 2 in. It goes through a hot-press machine that reduces the height of the cylinder to $\frac{1}{4}$ in. What is the new radius of the cylinder, assuming no material is lost?

10. Jack is making a scaled drawing of the floor plan of his home. The scale factor is 1 in:4 ft. The drawing of his living room is a rectangle measuring 5 inches by 3 inches. He is planning to purchase new carpet for the living room that costs $4 per square foot. How much will the carpet cost?
    (1) $720          (3) $960
    (2) $840          (4) $1,040

11. The city of Deersfield sits along the bank of the Fox Run River. A map of the city is shown below. The downtown region, indicated by region A, is $\frac{1}{2}$ mile wide and 1 mile long. The population density of Deersfield, except for the downtown region, is 800 people per square mile. The population density of the downtown region is 4,000 people per square mile. What is the population of Deersfield? Round to the nearest whole number.

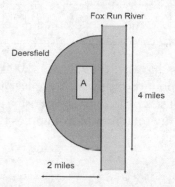

12. An engineer for a construction company needs to calculate the total volume of a home that is under construction so the proper sized heating and air conditioning equipment can be installed. The house is modeled as a rectangular prism with a triangular prism for the roof as shown in the accompanying figure. The roof is symmetric with $EF = ED$. The dimensions are $AB = 30$ ft, $BC = 50$ ft, $BD = 25$ ft, and the measure of $\angle EDF = 40°$. What is the total volume of the house? Round to the nearest cubic foot.

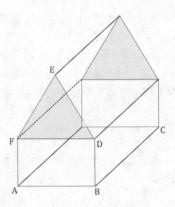

13. The Red Ribbon Orchard sorts their apple harvest by size. The three size categories are shown in the accompanying table.

|  | **Grade B** | **Grade A** | **Grade AA** |
|---|---|---|---|
| Average diameter | 3 inches | 4 inches | 5 inches |

After one day's harvest, there were 3,000 pounds of grade $B$ apples, 4,200 pounds of grade A apples, and 3,200 pounds of grade AA apples. If the average density of an apple is 0.016 lb/in³, what is the total number of apples that were harvested?

**14.** A construction company is working on plans for a project that call for digging a straight 0.25 mile tunnel through a mountain. The cross-section of the tunnel is shown in the accompanying figure.

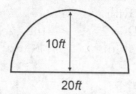

As the workers dig through the mountain, the rubble is brought to a gravel company to be processed into gravel. The cargo beds of the dump trucks are 18 ft long by 10 ft wide by 6 ft high, and the trucking company charges $450 per load delivered to the gravel company. Approximately how much money should the engineer budget for trucking costs for the project?

**15.** An engineer is designing a hinge to be used on the landing gear of a new airplane. She sketches the cross-section of the hinge on a computer, which is shown in the figure below. The circle represents a hole for a hinge pin.

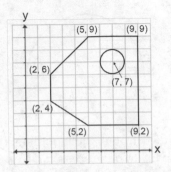

The engineer next uses a function on the computer that creates a solid by translating the cross-section in a direction perpendicular to the sketch. The engineer enters a translation distance of 10 units. In the final step, the engineer enters a scale factor of 1 unit = 2 centimeters. The resulting computer image of the solid is then sent to a factory to be manufactured out of a high-strength metal. If the density of the metal is 8.44 grams/cm$^3$ and the metal costs \$32 per kilogram, what is the expected cost for the metal to make one hinge? Round to the nearest dollar.

## Solutions

**1.** The basketball can be modeled as a sphere whose volume is $\frac{4}{3}\pi R^3$. The radius equals half the diameter, or 4.8 in.

$$V = \frac{4}{3}\pi R^3$$

$$= \frac{4}{3}\pi(4.8 \text{ in})^3$$

$$= 463 \text{ in}^3$$

The correct choice is **(3)**.

**2.** Start with the formula for volume, substitute the known values, and then solve for the radius.

$$V = B \cdot h$$
$$V = \pi R^2 h$$
$$200 = \pi(R^2)(6)$$
$$R^2 = 10.61032$$
$$R = \sqrt{10.61032}$$
$$= 3.3 \text{ cm}$$

The correct choice is **(3)**.

**3.** To find the weight of the topsoil, we need to first find the volume of the pile. The volume of a cone is $\frac{1}{3}Bh$, where $B$ is the area of the base. This requires finding the area of the circular base.

$$B = \pi R^2$$
$$= \pi(9^2)$$
$$= 254.4690 \text{ ft}^2$$

We are now ready to find the volume.

$$V = \frac{1}{3}B \cdot h$$

$$= \frac{1}{3}(254.4690)(3)$$

$$= 254.469 \text{ ft}^3$$

Now calculate the weight using the volume and density.

$$\begin{aligned}
\text{weight} &= V \cdot \text{density} \\
&= 254.469 \text{ ft}^3 \cdot 95 \text{ lb/ft}^3 \\
&= 24{,}174.55 \text{ lb} \\
&= 24{,}175 \text{ lb}
\end{aligned}$$

The correct choice is **(2)**.

**4.** The cross-section will be congruent to the base, so it is a circle with radius of 4. The area is $\pi R^2$, or $16\pi$.

The correct choice is **(4)**.

**5.** The strategy is to set the two volume formulas equal to each other and rearrange, solving for the ratio of the radii.

$$V_{\text{cone}} = \frac{1}{3}\pi R^2_{\text{cone}}$$

$$V_{\text{cylinder}} = \pi R^2_{\text{cylinder}}\, h$$

$$V_{\text{cylinder}} = V_{\text{cone}}$$

$$\pi R^2_{\text{cylinder}}\, h = \frac{1}{3}\pi R^2_{\text{cone}}\, h$$

Now solve for the ratio $\dfrac{R_{\text{cylinder}}}{R_{\text{cone}}}$

$$R^2_{\text{cylinder}} = \frac{1}{3}R^2_{\text{cone}} \qquad \text{divide by } \pi h$$

$$\frac{R^2_{\text{cylinder}}}{R^2_{\text{cone}}} = \frac{1}{3} \qquad \text{divide by } R^2_{\text{cone}}$$

$$\frac{R_{\text{cylinder}}}{R_{\text{cone}}} = \frac{1}{\sqrt{3}} \qquad \text{take square root of each side}$$

The ratio does not match any of the choices, so try rationalizing the denominator. To do so, multiply the numerator and denominator by $\sqrt{3}$.

$$\frac{R_{\text{cylinder}}}{R_{\text{cone}}} = \frac{1}{\sqrt{3}} \cdot \frac{\sqrt{3}}{\sqrt{3}}$$

$$\frac{R_{\text{cylinder}}}{R_{\text{cone}}} = \frac{\sqrt{3}}{3}$$

The correct choice is **(4)**.

**6.** A right triangle rotated 360° about one of its legs will generate a cone.

The leg aligned with the axis of rotation becomes the height, so the height equals 10. The leg perpendicular to the axis of rotation becomes the radius, so the radius equals 6 and the diameter equals 12.

The correct choice is **(3)**.

**7.** Cavalieri's principle states that two parallel planes will intercept the same volume in two solids if the cross-sectional areas are uniform and equal. Choice 3 represents this principle.

The correct choice is **(3)**.

**8.** $x^2 + y^2 = 16$ describes a circle centered at the origin with a radius of 4. Its area is

$$A = \pi R^2 = 16\pi$$

Apply the volume formula of the cone to find the height.

$$V = \frac{1}{3} Bh$$

$$72\pi = \frac{1}{3}16\pi h$$

$$h = 72 \cdot 3 \cdot \frac{1}{16}$$

$$= 13.5$$

**9.** Find the volume of the cylinder before pressing and set it equal to the expression for the volume after pressing. Use this equation to solve for the radius after pressing.

$$V = \pi R^2 h$$
$$V_{\text{before}} = \pi \cdot 15^2 \cdot 2$$
$$= 450\pi \text{ in}^3$$

$$V_{\text{after}} = \pi \cdot R^2 \cdot \frac{1}{4} \qquad \text{we don't know the new radius}$$

$$V_{\text{before}} = V_{\text{after}}$$

$$\pi \cdot R^2 \cdot \frac{1}{4} = 450\pi \qquad \text{volumes must be equal}$$

$$\frac{R^2}{4} = 450$$

$$R^2 = 1{,}800$$

$$R = \sqrt{1{,}800}$$

$$= 42.426 \text{ in}$$

**10.** The actual length and width of the living room are found by applying the scale factor to the drawing dimensions.

$$5 \text{ inches} \cdot \frac{4 \text{ ft}}{1 \text{ in}} = 20 \text{ ft}$$

$$3 \text{ inches} \cdot \frac{4 \text{ ft}}{1 \text{ in}} = 12 \text{ ft}$$

The area of the rectangular living room is

$$
\begin{aligned}
A &= \text{length} \cdot \text{width} \\
&= 20 \text{ ft} \cdot 12 \text{ ft} \\
&= 240 \text{ ft}^2
\end{aligned}
$$

The cost of the carpet is

$$\text{cost} = \text{area} \cdot \text{cost per ft}^2$$

$$\text{cost} = 240 \text{ ft}^2 \cdot \frac{\$4}{\text{ft}^2}$$

$$= \$960$$

The correct choice is **(3)**.

**11.** We can model the town as a semicircle and the downtown region as a rectangle. To calculate the population, we first find the area of each region. Let the area of the downtown region be represented by $A_A$ and the area of the remainder of the town be represented by $A_B$.

$$A_A = \text{length} \cdot \text{width} \qquad \text{area of circle}$$

$$= \frac{1}{2} \cdot 1$$

$$= 0.5 \text{ mi}^2$$

The area of the remainder of the town is a semicircle minus the area of the downtown region.

$$A_B = \frac{1}{2}\pi R^2 - 0.5 \qquad \text{area of semicircle circle – area}$$
$$\text{of rectangle}$$

$$= \frac{1}{2}\pi(2)^2 - 0.5$$

$$= 5.78318 \text{ mi}^2$$

The population of each region is the product of the area and the population density

Population $= A_A \cdot 800 \text{ people/mi}^2 + A_B \cdot 4{,}000 \text{ people/mi}^2$

$$= 5.78318 \text{ mi}^2 \cdot 800 \text{ people/mi}^2$$
$$+ 0.5 \text{ mi}^2 \cdot 4{,}000 \text{ people/mi}^2$$

$$= 6{,}627 \text{ people.}$$

The population of Deerfield is 6,627 people.

**12.** Calculate the volume of each section.
Rectangular prism:

length $= BC$      width $= AB$      height $= BD$
length $= 50$ ft    width $= 30$ ft    height $= 25$ ft

Volume$_{\text{rectangular prism}}$ = length $\cdot$ width $\cdot$ height
$$= (50 \text{ ft}) (30 \text{ ft}) (25 \text{ ft})$$
$$= 37{,}500 \text{ ft}^3$$

Triangular prism:

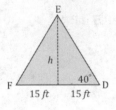

The first step is to find the area of the triangular base, $\triangle FED$. $FD$ has a length equal to $AB$, or 30 ft. The height of the isosceles triangle divides the base into two 15 ft lengths. The tangent ratio can be used to find the height of $\triangle FED$.

$$\tan = \frac{\text{opposite}}{\text{adjacent}}$$

$$\tan(40°) = \frac{h}{15}$$
$$h = 15 \tan(40°)$$
$$= 12.586 \text{ ft}$$

Now multiply the formula for the area of a triangle.

$$A_{\triangle FED} = 2 \cdot \frac{1}{2} \text{base} \cdot \text{height}$$
$$= 15 \text{ ft} \cdot 12.586 \text{ ft}$$
$$= 188.79 \text{ ft}^2$$

The volume of a triangular prism is equal to $Bh$, where $B$ is the area of the triangular base and $h$ is the length of the prism, which is equal to $BC$.

$$V_{\text{triangular prism}} = Bh$$
$$= A_{\triangle FED} \cdot BC$$
$$= 188.79 \text{ ft}^2 \cdot 50 \text{ ft}$$
$$= 9{,}439.5 \text{ ft}^3$$

Sum the two volumes to get the total volume of the house.

$$\text{Volume} = 37{,}500 \text{ ft}^3 + 9{,}439.5 \text{ ft}^3$$
$$= 46{,}939.5 \text{ ft}^3$$
$$= 46{,}940 \text{ ft}^3$$

**13.** The number of apples of each type equals the total volume of that type divided by the volume of one apple. Model the apples as spheres to calculate the volume of one apple. Find the total volume by dividing the weight by the density. The following table summarizes the calculations.

| | Grade B | Grade A | Grade AA |
|---|---|---|---|
| volume of 1 apple = $V = \frac{4}{3}\pi R^3$ | $= \frac{4}{3}\pi \cdot (1.5)^3$ $= 14.1372 \text{ in}^3$ | $= \frac{4}{3}\pi \cdot (2)^3$ $= 33.5103 \text{ in}^3$ | $= \frac{4}{3}\pi \cdot (2.5)^3$ $= 65.4498 \text{ in}^3$ |
| Total volume of each type = $\frac{\text{weight}}{\text{density}}$ | $= \frac{3,000 \text{ lb}}{0.016 \text{ lb/in}^3}$ $= 187,500 \text{ in}^3$ | $= \frac{4,200 \text{ lb}}{0.016 \text{ lb/in}^3}$ $= 262,500 \text{ in}^3$ | $= \frac{3,000 \text{ lb}}{0.016 \text{ lb/in}^3}$ $= 200,000 \text{ in}^3$ |
| number of each type = $\frac{V_{\text{total}}}{V_{\text{one apple}}}$ | $= \frac{187,500}{14.1372}$ $= 13,263$ | $= \frac{262,500}{33.5013}$ $= 7,833$ | $= \frac{200,000}{65.4498}$ $= 3,056$ |

The total number of apples is $13,263 + 7,833 + 3,056 = 24,152$ apples.

**14.** First, find the volume of the tunnel, and divide by the volume of the dump truck bed to find the number of dump-truck loads needed. The cross-section of the tunnel is a semicircle, which makes the tunnel in the shape of a half cylinder. The cylinder radius is 10 ft and the length is 0.25 miles.

First, convert 0.25 miles to feet using the conversion factor from the formula sheet.

$$0.25 \text{ mi} \cdot \frac{5{,}280 \text{ ft}}{\text{mi}} = 1{,}320 \text{ ft}$$

$$V_{\text{tunnel}} = \frac{1}{2}\pi R^2 h$$

$$= \frac{1}{2}\pi \cdot 10^2 \cdot 1{,}320$$

$$= 207{,}345.11 \text{ ft}^3$$

Model the dump-truck bed as a rectangular prism.

$$V_{\text{truck}} = \text{length} \cdot \text{width} \cdot \text{height}$$
$$= 18 \cdot 10 \cdot 6 = 1{,}080 \text{ ft}^3$$

The number of dump-truck loads is the ratio of the two volumes.

$$\text{Number of loads} = V_{\text{tunnel}} / V_{\text{truck}}$$

$$= \frac{207{,}345.11}{1{,}080}$$

$$= 191.98 \text{ dump-truck loads}$$

192 dump-truck loads are needed.

$$\text{The estimated cost} = \frac{\$450}{\text{load}} \cdot 192 \text{ loads} = \$86{,}400$$

**15.** The solid produced will be a prism with the cross-section shown in the figure. The first step is to calculate the area of the cross-section. Divide the cross-section into the 5 regions shown below. The 5th region represents the hole and must be subtracted from the other areas.

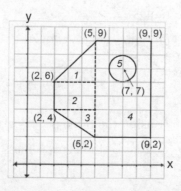

| $A_1 = \frac{1}{2}b \cdot h$ | $A_2 = l \cdot w$ | $A_3 = \frac{1}{2}b \cdot h$ | $A_4 = l \cdot w$ | $A_5 = \pi R^2$ |
|---|---|---|---|---|
| $= \frac{1}{2} \cdot 3 \cdot 3$ | $= 3 \cdot 2$ | $= \frac{1}{2} \cdot 3 \cdot 2$ | $= 4 \cdot 7$ | $= \pi(1)^2$ |
| $= 4.5$ | $= 6$ | $= 3$ | $= 28$ | $= 3.14159$ |

$$A_{\text{drawing}} = A_1 + A_2 + A_3 + A_4 - A_5$$
$$= 4.5 + 6 + 3 + 28 - 3.14159 = 38.35841 \text{ units}^2$$

To find the cross-sectional area of the actual part in centimeters, use the scale factor and the fact that area is proportional to the scale factor *squared*.

$$A_{\text{cross-section}} = 38.35841 \text{ units}^2 \cdot \left(\frac{2 \text{ cm}}{1 \text{ unit}}\right)^2$$
$$= 153.43364 \text{ cm}^2$$

The solid is formed by translating the cross-section 10 units, or 20 cm. Use these dimensions to find the volume of the prism.

$V = B \times h$        where $B$ is the cross-sectional area
$= 153.43364 \text{ cm}^2 \cdot 20 \text{ cm}$
$= 3{,}068.6728 \text{ cm}^3$

Now calculate the mass of the metal

$= 8.44 \dfrac{\text{grams}}{\text{cm}^3} \cdot 3{,}068.6728 \text{ cm}^3$     mass = density × volume
$= 25{,}899.5984 \text{ grams}$

The price is given per kilogram, so convert grams to kilograms.

$\text{Mass} = 25{,}899.5984 \text{ grams} \cdot \dfrac{1 \text{ kg}}{1{,}000 \text{ grams}}$
$= 25.899 \text{ kg}$

Finally, find the cost using the price per kilogram.

$\text{Cost} = 25.899 \text{ kg} \cdot \$32/\text{kg}$     total cost = mass (price per kg)
$= \$829$

# Glossary of Geometry Terms

**Acute angle** An angle whose measure is greater than 0° and less than 90°.

**Acute triangle** A triangle whose angles are all acute angles.

**Adjacent angles** Two angles that share a common vertex and one side, but do not share interior points.

**Altitude (of a triangle)** A segment from a vertex of a triangle perpendicular to the opposite side.

**Angle** A figure formed by two rays with a common endpoint. Symbol is ∠.

**Angle bisector** A line, segment, or ray that divides an angle into two congruent angles.

**Angle measure** The amount of opening of an angle, measured in degrees or radians.

**Angle of depression** The angle formed by the horizontal and the line of sight when looking downward to an object.

**Angle of elevation** The angle formed by the horizontal and the line of sight when looking upward to an object.

**Angle of rotation** The angle measure by which a figure or point spins around a center point.

**Arc** A portion of a circle, with two endpoints on the circle. A **major arc** measures more than 180°, a **minor arc** measures less than 180°, and a **semicircle** measures 180°. Symbol ⌒.

**Arc length** The distance between the endpoints of an arc. Arc length equals $R\theta$ where $\theta$ is the arc measure in radians and $R$ is the radius.

**Arc measure** The angle measure of an arc equal to the measure of the central angle that intercepts the arc.

**Base (of a circular cone)** The circular face of a cone; it is opposite the apex.

**Base (of an isosceles triangle)** The noncongruent side of an isosceles triangle.

**Base (of a pyramid)** The polygonal face of a pyramid that is opposite the apex.

**Bases (of a prism)** A pair of faces of a prism that are parallel, congruent polygons.

**Bases (of a circular cylinder)** The pair of congruent parallel circular faces of a cylinder.

**Base angles** The angles formed at each end of the base of an isosceles triangle.

**Bisector**, or **segment bisector** A line, segment, or ray that passes through the midpoint of a segment.

**Biconditional** A compound statement in the form "If and only if *hypothesis* then *conclusion*." The hypothesis and conclusion are statements that can be true or false. The biconditional is true when the hypothesis and conclusion have the same truth value.

**Cavalieri's principle** If two solids are contained between two parallel planes, and every parallel plane between these two planes intercepts regions of equal area, then the solids have equal volume. Also, any two parallel planes intercept two solids of equal volume.

**Center of dilation** The fixed reference point used to determine the expansion or contraction of lengths in a dilation. The center of dilation is the only invariant point in a dilation.

**Center of a regular polygon** The center of the inscribed or circumscribed circle of a regular polygon.

**Center–radius equation of a circle** A circle can be represented on the coordinate plane by $(x - h)^2 + (y - k)^2 = r^2$, where the point $(h, k)$ is the center of the circle and $r$ is the radius.

**Central angle** An angle in a circle formed by two distinct radii.

**Centroid of a triangle** The point of concurrency of the three medians of a triangle.

**Chord** A segment whose endpoints lie on a circle.

**Circle** The set of points that are a fixed distance from a fixed center point. Symbol ⊙.

**Circumcenter** The point that is the center of the circle circumscribed about a polygon, equidistant from the vertices of a polygon, and the point of concurrency of the perpendicular bisectors of a triangle.

**Circumference** The distance around a circle. Circumference = $2\pi \cdot$ radius.

**Coincide (coincident)** Figures that lay entirely one on the other.

**Collinear** Points that lie on the same line.

**Compass** A tool for drawing accurate circles and arcs.

**Complementary angles** Angles whose measures sum to 90°.

**Completing the square** A method used to rewrite a quadratic expression of the form $ax^2 + bx + c$ as a squared binomial of the form $(dx + e)^2$.

**Composition of transformations** A sequence of transformations in a specified order.

**Concave polygon** A polygon with at least one diagonal outside the polygon.

**Concentric circles** Circles with the same center.

**Concurrent** When three or more lines all intersect at a single point.

**Cone (circular)** A solid figure with a single circular base, and a curved lateral face that tapers to a point called the apex.

**Congruent** Figures with the same size and shape. Symbol ≅.

**Construction** A figure drawn with only a compass and straightedge.

**Convex polygon** A polygon whose diagonals all lie within the polygon.

**Coplanar** Figures that lie in the same plane.

**Corresponding parts** A pair of parts (usually points, sides, or angles) of two figures that are paired together through a specified relationship, such as a congruence or similarity statement or a transformation function.

**Cosine of an angle** In a right triangle, the ratio of the length of the side adjacent to an acute angle to the length of the hypotenuse.

**CPCTC** Corresponding parts of congruent triangles are congruent.

**Cross-section** The intersection of a plane with a solid.

**Cube** A prism whose faces are all squares.

**Cubic unit** The amount of space occupied by a cube 1 unit of length in each dimension. (1 ft$^3$ is the amount of space occupied by a 1 ft × 1 ft × 1 ft cube.)

**Cylinder** The solid figure formed when a rectangle is rotated 360° about one of its sides.

**Cylinder (circular)** A solid with two parallel, congruent circular bases and a curved lateral face.

**Degree** A unit of angle measure equal to $\frac{1}{360}$ th of a complete rotation.

**Diagonal** A segment in a polygon whose endpoints are nonconsecutive vertices of the polygon.

**Diameter** A chord that passes through the center of a circle.

**Dilation** A similarity transformation about a center point $O$, which maps pre-image $P$ to image $P'$ such that $O$, $P$, and $P'$ are collinear and $OP' = k \cdot OP$. Notation is $D_k$.

**Direct transformation** A transformation that preserves orientation.

**Distance from a point to a line** The length of the segment from the point perpendicular to the line.

**Edge of a solid** The intersection of two polygonal faces of a solid.

**Endpoint** The two points that bound a segment.

**Equiangular** A figure whose angles all have the same measure.

**Equidistant** The same distance from two or more points.

**Equilateral** A figure whose sides all have the same length.

**Exterior angle of a polygon** The angle formed by a side of a polygon and the extension of an adjacent side.

**Face of a solid** Any of the surfaces that bound a solid.

**Figure** Any set of points. The points may form a segment, line, ray, plane, polygon, curve, solid, etc.

**Glide reflection** The composition of a line reflection and a translation along a vector parallel to the line of reflection.

**Great circle** The largest circle that can be drawn on a sphere. The circle formed by the intersection of a sphere and a plane passing through the center of the sphere.

**Hexagon** A 6-sided polygon.

**Hypotenuse** The longest side of a right triangle; it is always opposite the right angle.

**Identity transformation** A transformation in which the pre-image and image are coincident.

**Image** The figure that results from applying a transformation to an initial figure called the pre-image.

**Incenter of a triangle** The point that is the center of the inscribed circle of a triangle, equidistant from the 3 sides of a triangle, and the point of concurrency of the three angle bisectors of a triangle.

**Inscribed angle** An angle in a circle formed by two chords with a common endpoint.

**Inscribed circle of a triangle** A circle that is tangent to all three sides of the triangle.

**Intersecting** Figures that share at least one common point.

**Isometry** See *Rigid motion.*

**Isosceles trapezoid** A trapezoid with congruent legs.

**Isosceles triangle** A triangle with at least two congruent sides.

**Kite** A polygon with two distinct pairs of adjacent congruent sides. The opposite sides are not congruent.

**Lateral edge** The intersection between two lateral faces of a polyhedron.

**Lateral face** Any face of a polyhedron that is not a base.

**Line** One of the undefined terms in geometry. An infinitely long set of points that has no width or thickness. Symbol $\leftrightarrow$.

**Linear pair of angles** Two adjacent supplementary angles.

**Line symmetry** A line over which one half of a figure can be reflected and mapped onto the other half of the figure.

**Line reflection** A rigid motion in which every point $P$ is transformed to a point $P'$ such that the line is the perpendicular bisector of $\overline{PP'}$.

**Major arc** An arc whose degree measure is greater than 180°.

**Map (mapping)** A pairing of every point in a pre-image with a point in an image. Reflections, rotations, translations, and dilations are one-to-one mappings because every point in the pre-image maps to exactly one point in the image, and every point in the image maps to exactly one point in the pre-image.

**Mean proportional (geometric mean)** The square root of the product of two numbers, $a$ and $b$. If $\dfrac{a}{m} = \dfrac{m}{b}$, then $m$ is the geometric mean.

**Median of a triangle** The segment from a vertex of a triangle to the midpoint of the opposite side.

**Midpoint** A point that divides a segment into two congruent segments.

**Midsegment (median) of a trapezoid** A segment joining the midpoints of the two legs of a trapezoid.

**Midsegment of a triangle** A segment joining the midpoints of two sides of a triangle.

**Minor arc** An arc whose degree measure is less than 180°.

**Noncollinear** Points that do not lie on the same line.

**Noncoplanar** Points or lines that do not lie on the same plane.

**Obtuse angle** An angle whose measure is greater than 90° and less than 180°.

**Octagon** A polygon with 8 sides.

**Opposite rays** Two rays with a common endpoint that together form a straight line.

**Opposite transformation** A transformation that changes the orientation of a figure.

**Orientation** The order in which the points of a figure are encountered when moving around a figure. The orientation can be clockwise or counterclockwise.

**Orthocenter** The point of concurrence of the three altitudes of a triangle.

**Parallel lines** Coplanar lines that do not intersect. Symbol //.

**Parallelogram** A quadrilateral with two pairs of opposite parallel sides.

**Parallel planes** Planes that do not intersect.

**Pentagon** A polygon with 5 sides.

**Perimeter** The sum of the lengths of the sides of a polygon.

**Perpendicular** Intersecting at right angles. Symbol ⊥.

**Perpendicular bisector** A line, segment, or ray perpendicular to another segment at its midpoint.

**Pi** The ratio of the circumference of a circle to its diameter. Pi is an irrational number whose value is approximately 3.14159. Symbol π.

**Plane** One of the undefined terms in geometry. A set of points with no thickness that extends infinitely in all directions. It is usually visualized as a flat surface.

**Point** An undefined term in geometry. A location in space with no length, width, or thickness. Symbol •.

**Point of tangency** The point where a tangent intersects a curve.

**Point reflection** A transformation in which a specified center point is the midpoint of each of the segments connecting any point in the pre-image with its corresponding point in the image. It is equivalent to a rotation of 180°.

**Point-slope equation of a line** A form of the equation of a line, $y - y_1 = m(x - x_1)$, where $m$ is the slope and $(x_1, y_1)$ are the coordinates of a point on the line.

**Polygon** A closed planar figure whose sides are segments that intersect only at their endpoints (do not overlap).

**Polyhedron** A solid figure in which each face is a polygon. Plural *polyhedra*. Prisms and pyramids are examples of polyhedra.

**Postulate** A statement that is accepted to be true without proof.

**Pre-image** The original figure that is acted on by a transformation.

**Preserves** Remains unchanged.

**Prism** A polyhedron with two congruent, parallel, polygons for bases, and whose lateral faces are parallelograms.

**Proportion** An equation that states two ratios are equal. For example, $\frac{a}{b} = \frac{c}{d}$. The cross-products of a proportion are equal, $a \cdot d = b \cdot c$.

**Pyramid** A polyhedron having one polygonal base and triangles for lateral faces.

**Pythagorean theorem** In a right triangle, the sum of the squares of the two legs equals the square of the hypotenuse, or $a^2 + b^2 = c^2$, where $a$ and $b$ are the lengths of the two legs and $c$ is the length of the hypotenuse.

**Quadratic equation** An equation in the form $ax^2 + bx + c = 0$, where $a \neq 0$.

**Quadratic formula** A formula for finding the two solutions to a quadratic equation of the form $ax^2 + bx + c = 0$, $x = \dfrac{-b \pm \sqrt{b^2 - 4ac}}{2a}$.

**Quadrilateral** A polygon with four sides.

**Radian** An angle measure in which one full rotation is $2\pi$ radians. Also, 1 radian is the measure of an arc such that the arc's length is equal to the radius of that circle.

**Radius** A segment from the center of a circle to a point on the circle.

**Ray** A portion of a line starting at an endpoint and including all points on one side of the endpoint. Symbol $\rightarrow$.

**Rectangle** A parallelogram with right angles.

**Reflection** See *Line reflection* and *Point reflection*.

**Reflexive property of equality** Any quantity is equal to itself. Also, for figures, any figure is congruent to itself.

**Regular polygon** A polygon in which all sides are congruent and all angles are congruent.

**Rhombus** A parallelogram with 4 congruent sides.

**Right angle** An angle that measures 90°.

**Right circular cone** A cone with a circular base and whose altitude passes through the center of the base.

**Right circular cylinder** A cylinder with circular bases and whose altitude passes through the center of the bases.

**Right pyramid** A pyramid whose faces are isosceles triangles.

**Right triangle** A triangle that contains a right angle.

**Rigid motion (isometry)** A transformation that preserves distance. The image and pre-image are congruent under a rigid motion. Translations, reflections, and rotations are isometries.

**Rotation** A rigid motion in which every point, $P$, in the pre-image spins by a fixed angle around a center point, $C$, to point $P'$. The distance to the center point is preserved.

**Rotational symmetry** A figure has rotational symmetry if it maps onto itself after a rotation of less than 360°.

**Scale drawing** A drawing of a figure or object that represents a dilation of the actual figure or object. A drawing of an object in which every length is enlarged or reduced by the same scale factor.

**Scale factor** The ratio by which a figure is enlarged or reduced by a dilation.

**Scalene triangle** A triangle in which no sides have the same length.

**Secant of a circle** A line that intersects a circle in exactly two points.

**Sector of a circle** A region bounded by a central angle of a circle and the arc it intersects.

**Segment** A portion of a line bounded by two endpoints.

**Semicircle** An arc that measures 180°.

**Similarity transformation** A transformation in which the pre-image and image are similar. A transformation that includes a dilation.

**Similar polygons** Two polygons with the same shape but not necessarily the same size.

**Sine of an angle** In a right triangle, the ratio of the length of the side opposite an acute angle to the length of the hypotenuse.

**Skew lines** Two lines that are not coplanar.

**Slant height** The distance along a lateral face of a solid from the apex perpendicular to the opposite edge.

**Slope** A numerical measure of the steepness of a line. In the coordinate plane, the slope of a line equals the change in the $y$-coordinates divided by the change in the $x$-coordinates between any two points. The slope of a vertical line is undefined.

**Slope-intercept equation of a line** A form of the equation of a line, $y = mx + b$, where $m$ is the slope and $b$ is the $y$-intercept.

**Solid figure** A 3-dimensional figure fully enclosed by surfaces.

**Sphere** A solid comprised of the set of points in space that are a fixed distance from a center point.

**Square** A parallelogram with 4 right angles and 4 congruent sides.

**Straightedge** A ruler with no length marking, used for constructing straight lines.

**Substitution property of equality** The property that a quantity can be replaced by an equal quantity in an equation.

**Subtraction property of equality** If the same or equal quantities are subtracted from the same or equal quantities, then the differences are equal.

**Supplementary angles** Two angles whose measures sum to 180°.

**Surface area** The sum of the areas of all the faces or curved surfaces of a solid figure.

**Tangent of an angle** In a right triangle, the ratio of the length of the side opposite to an acute angle to the length of the side adjacent to the angle.

**Tangent to a circle** A line, coplanar with a circle, that intersects the circle at only one point.

**Theorem** A general statement that can be proven.

**Transformation** A one-to-one function that maps a set of points, called the pre-image, to a new set of points, called the image.

**Transitive property of congruence** The property that states: if figure $A \cong$ figure $B$ and figure $B \cong$ figure $C$, then figure $A \cong$ figure $C$.

**Transitive property of equality** The property that states: if $a = b$ and $b = c$, then $a = c$.

**Translation** A rigid motion that slides every point in the pre-image in the same direction and by the same distance.

**Transversal** A line that intersects two or more other lines at different points.

**Trapezoid** A quadrilateral that has at least one pair of parallel sides.

**Undefined terms** Terms which cannot be formally defined using previously defined terms. In geometry these traditionally include point, line, and plane.

**Vector** A quantity that has both magnitude and direction; represented geometrically by a directed line segment. Symbol $\rightarrow$.

**Vertex** The point of intersection of two consecutive sides of a polygon or the two rays of an angle.

**Vertex angle** The angles formed by the two congruent sides in an isosceles triangle.

**Vertical angles** The pairs of opposite angles formed by two intersecting lines.

**Volume** The amount of space occupied by a solid figure measured in cubic units ($in^3$, $cm^3$, etc.). The number of nonoverlapping unit cubes that can fit in the interior of a solid.

**Zero product property** The property that states: if $a \cdot b = 0$, then either $a = 0$, or $b = 0$, or $a$ and $b = 0$.

# Regents Examinations, Answers, and Self-Analysis Charts

# Examination
# June 2015
## Geometry (Common Core)

## GEOMETRY REFERENCE SHEET

| | |
|---|---|
| 1 inch = 2.54 centimeters | 1 ton = 2000 pounds |
| 1 meter = 39.37 inches | 1 cup = 8 fluid ounces |
| 1 mile = 5280 feet | 1 pint = 2 cups |
| 1 mile = 1760 yards | 1 quart = 2 pints |
| 1 mile = 1.609 kilometers | 1 gallon = 4 quarts |
| 1 kilometer = 0.62 mile | 1 gallon = 3.785 liters |
| 1 pound = 16 ounces | 1 liter = 0.264 gallon |
| 1 pound = 0.454 kilogram | 1 liter = 1000 cubic centimeters |
| 1 kilogram = 2.2 pounds | |

| | |
|---|---|
| Triangle | $A = \frac{1}{2}bh$ |
| Parallelogram | $A = bh$ |
| Circle | $A = \pi r^2$ |
| Circle | $C = \pi d$ or $C = 2\pi r$ |
| General Prisms | $V = Bh$ |
| Cylinder | $V = \pi r^2 h$ |
| Sphere | $V = \frac{4}{3}\pi r^3$ |

| Cone | $V = \frac{1}{3}\pi r^2 h$ |
|---|---|
| Pyramid | $V = \frac{1}{3}Bh$ |
| Pythagorean Theorem | $a^2 + b^2 = c^2$ |
| Quadratic Formula | $x = \dfrac{-b \pm \sqrt{b^2 - 4ac}}{2a}$ |
| Arithmetic Sequence | $a_n = a_1 + (n-1)d$ |
| Geometric Sequence | $a_n = a_1 r^{n-1}$ |
| Geometric Series | $S_n = \dfrac{a_1 - a_1 r^n}{1 - r}$ where $r \neq 1$ |
| Radians | 1 radian $= \dfrac{180}{\pi}$ degrees |
| Degrees | 1 degree $= \dfrac{\pi}{180}$ radians |
| Exponential Growth/Decay | $A = A_0 e^{k(t - t_0)} + B_0$ |

## PART I

Answer all 24 questions in this part. Each correct answer will receive 2 credits. No partial credit will be allowed. For each statement or question, write in the space provided the numeral preceding the word or expression that best completes the statement or answers the question. [48 credits]

1 Which object is formed when right triangle *RST* shown below is rotated around leg $\overline{RS}$?

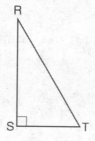

   (1) a pyramid with a square base
   (2) an isosceles triangle
   (3) a right triangle
   (4) a cone                 1 _____

2 The vertices of △*JKL* have coordinates *J*(5, 1), *K*(2, 3), and *L*(4, 1). Under which transformation is the image △*J'K'L'* *not* congruent to △*JKL*?

(1) a translation of two units to the right and two units down

(2) a counterclockwise rotation of 180 degrees around the origin

(3) a reflection over the *x*-axis

(4) a dilation with a scale factor of 2 and centered at the origin                                                          2 _____

3 The center of circle *Q* has coordinates (3, –2). If circle *Q* passes through *R*(7, 1), what is the length of its diameter?

(1) 50                     (3) 10

(2) 25                     (4) 5                                 3 _____

**4** In the diagram below, congruent figures 1, 2, and 3 are drawn.

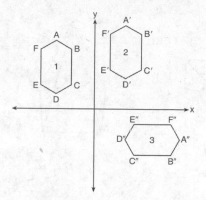

Which sequence of transformations maps figure 1 onto figure 2 and then figure 2 onto figure 3?

(1) a reflection followed by a translation
(2) a rotation followed by a translation
(3) a translation followed by a reflection
(4) a translation followed by a rotation      4 _____

**5** As shown in the diagram below, the angle of elevation from a point on the ground to the top of the tree is 34°.

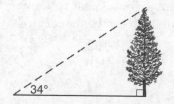

If the point is 20 feet from the base of the tree, what is the height of the tree, to the *nearest tenth of a foot*?

(1) 29.7          (3) 13.5
(2) 16.6          (4) 11.2      5 _____

**6** Which figure can have the same cross-section as a sphere?

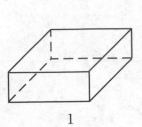

1

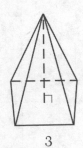

3

2

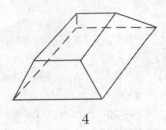

4

6 ____

**7** A shipping container is in the shape of a right rectangular prism with a length of 12 feet, a width of 8.5 feet, and a height of 4 feet. The container is completely filled with contents that weigh, on average, 0.25 pound per cubic foot. What is the weight, in pounds, of the contents in the container?

(1) 1,632        (3) 102

(2) 408          (4) 92

7 ____

8 In the diagram of circle $A$ shown below, chords $\overline{CD}$ and $\overline{EF}$ intersect at $G$, and chords $\overline{CE}$ and $\overline{FD}$ are drawn.

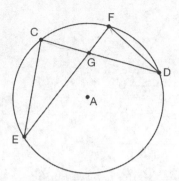

Which statement is *not* always true?

(1) $\overline{CG} \cong \overline{FG}$          (3) $\dfrac{CE}{EG} = \dfrac{FD}{DG}$

(2) $\angle CEG \cong \angle FDG$          (4) $\triangle CEG \sim \triangle FDG$          8 _____

9 Which equation represents a line that is perpendicular to the line represented by $2x - y = 7$?

(1) $y = -\dfrac{1}{2}x + 6$          (3) $y = -2x + 6$

(2) $y = \dfrac{1}{2}x + 6$          (4) $y = 2x + 6$          9 _____

10 Which regular polygon has a minimum rotation of $45°$ to carry the polygon onto itself?

(1) octagon          (3) hexagon
(2) decagon          (4) pentagon          10 _____

**11** In the diagram of $\triangle ADC$ below, $\overline{EB} \parallel \overline{DC}$, $AE = 9$, $ED = 5$, and $AB = 9.2$.

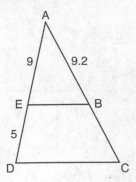

What is the length of $\overline{AC}$, to the *nearest tenth*?

(1) 5.1          (3) 14.3
(2) 5.2          (4) 14.4          11 _____

**12** In scalene triangle $ABC$ shown in the diagram below, $m\angle C = 90°$.

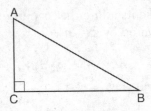

Which equation is always true?

(1) $\sin A = \sin B$          (3) $\cos A = \sin C$
(2) $\cos A = \cos B$          (4) $\sin A = \cos B$          12 _____

**13** Quadrilateral *ABCD* has diagonals $\overline{AC}$ and $\overline{BD}$. Which information is not sufficient to prove *ABCD* is a parallelogram?

(1) $\overline{AC}$ and $\overline{BD}$ bisect each other.

(2) $\overline{AB} \cong \overline{CD}$ and $\overline{BC} \cong \overline{AD}$

(3) $\overline{AB} \cong \overline{CD}$ and $\overline{AB} \parallel \overline{CD}$

(4) $\overline{AB} \cong \overline{CD}$ and $\overline{BC} \parallel \overline{AD}$

13 _____

**14** The equation of a circle is $x^2 + y^2 + 6y = 7$. What are the coordinates of the center and the length of the radius of the circle?

(1) center (0, 3) and radius 4

(2) center (0, –3) and radius 4

(3) center (0, 3) and radius 16

(4) center (0, –3) and radius 16

14 _____

**15** Triangles *ABC* and *DEF* are drawn below.

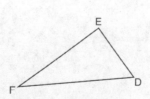

If *AB* = 9, *BC* = 15, *DE* = 6, *EF* = 10, and $\angle B \cong \angle E$, which statement is true?

(1) $\triangle CAB \cong \triangle DEF$

(3) $\triangle ABC \sim \triangle DEF$

(2) $\dfrac{AB}{CB} = \dfrac{FE}{DE}$

(4) $\dfrac{AB}{DE} = \dfrac{FE}{CB}$

15 _____

**16** If △ABC is dilated by a scale factor of 3, which statement is true of the image △A'B'C'?

(1) $3A'B' = AB$

(2) $B'C' = 3BC$

(3) $m\angle A' = 3(m\angle A)$

(4) $3(m\angle C') = m\angle C$

16 _____

**17** Steve drew line segments ABCD, EFG, BF, and CF as shown in the diagram below. Scalene △BFC is formed.

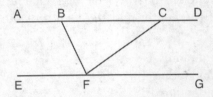

Which statement will allow Steve to prove $\overline{ABCD} \parallel \overline{EFG}$?

(1) $\angle CFG \cong \angle FCB$

(2) $\angle ABF \cong \angle BFC$

(3) $\angle EFB \cong \angle CFB$

(4) $\angle CBF \cong \angle GFC$

17 _____

**18** In the diagram below, $\overline{CD}$ is the image of $\overline{AB}$ after a dilation of scale factor $k$ with center $E$.

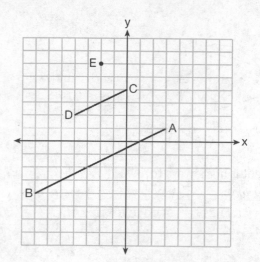

Which ratio is equal to the scale factor $k$ of the dilation?

(1) $\dfrac{EC}{EA}$      (3) $\dfrac{EA}{BA}$

(2) $\dfrac{BA}{EA}$      (4) $\dfrac{EA}{EC}$      18 _____

**19** A gallon of paint will cover approximately 450 square feet. An artist wants to paint all the outside surfaces of a cube measuring 12 feet on each edge. What is the *least* number of gallons of paint he must buy to paint the cube?

(1) 1      (3) 3

(2) 2      (4) 4      19 _____

**20** In circle $O$ shown below, diameter $\overline{AC}$ is perpendicular to $\overline{CD}$ at point $C$, and chords $\overline{AB}$, $\overline{BC}$, $\overline{AE}$, and $\overline{CE}$ are drawn.

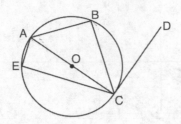

Which statement is *not* always true?

(1) $\angle ACB \cong \angle BCD$     (3) $\angle BAC \cong \angle DCB$

(2) $\angle ABC \cong \angle ACD$     (4) $\angle CBA \cong \angle AEC$     20 _____

**21** In the diagram below, $\triangle ABC \sim \triangle DEC$.

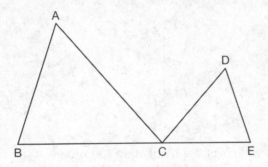

If $AC = 12$, $DC = 7$, $DE = 5$, and the perimeter of $\triangle ABC$ is 30, what is the perimeter of $\triangle DEC$?

(1) 12.5     (3) 14.8

(2) 14.0     (4) 17.5     21 _____

**22** The line $3y = -2x + 8$ is transformed by a dilation centered at the origin. Which linear equation could be its image?

(1) $2x + 3y = 5$        (3) $3x + 2y = 5$

(2) $2x - 3y = 5$        (4) $3x - 2y = 5$     22 _____

**23** A circle with a radius of 5 was divided into 24 congruent sectors. The sectors were then rearranged, as shown in the diagram below.

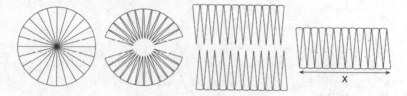

To the *nearest integer*, the value of $x$ is

(1) 31        (3) 12

(2) 16        (4) 10     23 _____

**24** Which statement is sufficient evidence that △DEF is congruent to △ABC?

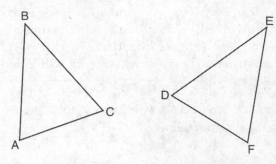

(1) $AB = DE$ and $BC = EF$

(2) $\angle D \cong \angle A, \angle B \cong \angle E, \angle C \cong \angle F$

(3) There is a sequence of rigid motions that maps $\overline{AB}$ onto $\overline{DE}, \overline{BC}$ onto $\overline{EF}$, and $\overline{AC}$ onto $\overline{DF}$.

(4) There is a sequence of rigid motions that maps point $A$ onto point $D, \overline{AB}$ onto $\overline{DE}$, and $\angle B$ onto $\angle E$.

24 _____

## PART II

Answer all 7 questions in this part. Each correct answer will receive 2 credits. Clearly indicate the necessary steps, including appropriate formula substitutions, diagrams, graphs, charts, etc. For all questions in this part, a correct numerical answer with no work shown will receive only 1 credit.   [14 credits]

**25** Use a compass and straightedge to construct an inscribed square in circle $T$ shown below.

    [Leave all construction marks.]

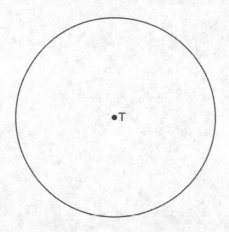

26 The diagram below shows parallelogram *LMNO* with
diagonal $\overline{LN}$, m∠*M* = 118°, and m∠*LNO* = 22°.

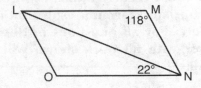

Explain why m∠*NLO* is 40 degrees.

**27** The coordinates of the endpoints of $\overline{AB}$ are $A(-6, -5)$ and $B(4, 0)$. Point $P$ is on $\overline{AB}$. Determine and state the coordinates of point $P$, such that $AP:PB$ is 2:3. [The use of the set of axes below is optional.]

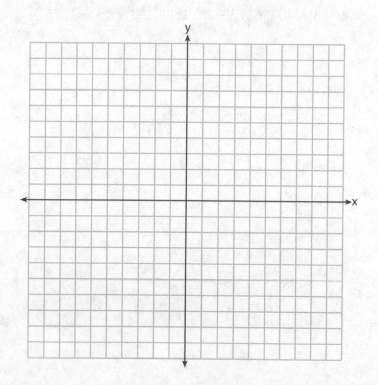

28 The diagram below shows a ramp connecting the ground to a loading platform 4.5 feet above the ground. The ramp measures 11.75 feet from the ground to the top of the loading platform.

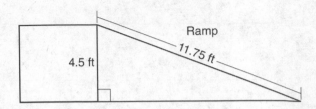

Determine and state, to the *nearest degree*, the angle of elevation formed by the ramp and the ground.

**29** In the diagram below of circle $O$, the area of the shaded sector $AOC$ is $12\pi$ in$^2$ and the length of $\overline{OA}$ is 6 inches. Determine and state m$\angle AOC$.

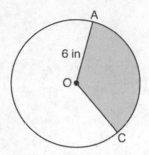

**30** After a reflection over a line, $\triangle A'B'C'$ is the image of $\triangle ABC$. Explain why triangle $ABC$ is congruent to triangle $A'B'C'$.

**31** A flagpole casts a shadow 16.60 meters long. Tim stands at a distance of 12.45 meters from the base of the flagpole, such that the end of Tim's shadow meets the end of the flagpole's shadow. If Tim is 1.65 meters tall, determine and state the height of the flagpole to the *nearest tenth of a meter.*

## PART III

Answer all 3 questions in this part. Each correct answer will receive 4 credits. Clearly indicate the necessary steps, including appropriate formula substitutions, diagrams, graphs, charts, etc. For all questions in this part, a correct numerical answer with no work shown will receive only 1 credit.   [12 credits]

32 In the diagram below, $\overline{EF}$ intersects $\overline{AB}$ and $\overline{CD}$ at $G$ and $H$, respectively, and $\overline{GI}$ is drawn such that $\overline{GH} \cong \overline{IH}$.

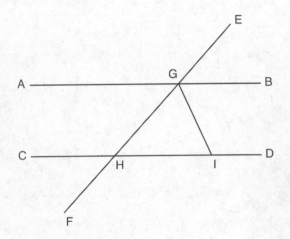

If m$\angle EGB$ = 50° and m$\angle DIG$ = 115°, explain why $\overline{AB} /\!/ \overline{CD}$.

**33** Given: Quadrilateral $ABCD$ is a parallelogram with diagonals $\overline{AC}$ and $\overline{BD}$ intersecting at $E$

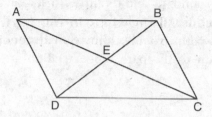

Prove: $\triangle AED \cong \triangle CEB$

Describe a single rigid motion that maps $\triangle AED$ onto $\triangle CEB$.

**34** In the diagram below, the line of sight from the park ranger station, $P$, to the lifeguard chair, $L$, on the beach of a lake is perpendicular to the path joining the campground, $C$, and the first aid station, $F$. The campground is 0.25 mile from the lifeguard chair. The straight paths from both the campground and first aid station to the park ranger station are perpendicular.

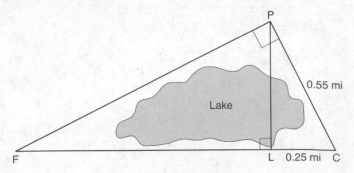

If the path from the park ranger station to the campground is 0.55 mile, determine and state, to the *nearest hundredth of a mile*, the distance between the park ranger station and the lifeguard chair.

Gerald believes the distance from the first aid station to the campground is at least 1.5 miles. Is Gerald correct? Justify your answer.

## PART IV

Answer the **2** questions in this part. Each correct answer will receive **6** credits. Clearly indicate the necessary steps, including appropriate formula substitutions, diagrams, graphs, charts, etc. For all questions in this part, a correct numerical answer with no work shown will receive only 1 credit. [12 credits]

**35** The water tower in the picture below is modeled by the two-dimensional figure beside it. The water tower is composed of a hemisphere, a cylinder, and a cone. Let *C* be the center of the hemisphere and let *D* be the center of the base of the cone.

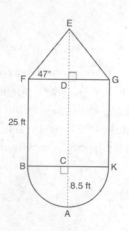

Source: http://en.wikipedia.org

**Question 35 is continued on the next page.**

## Question 35 continued

If $AC$ = 8.5 feet, $BF$ = 25 feet, and m$\angle EFD$ = 47°, determine and state, to the *nearest cubic foot*, the volume of the water tower.

The water tower was constructed to hold a maximum of 400,000 pounds of water. If water weighs 62.4 pounds per cubic foot, can the water tower be filled to 85% of its volume and not exceed the weight limit? Justify your answer.

**36** In the coordinate plane, the vertices of $\triangle RST$ are $R(6, -1)$, $S(1, -4)$, and $T(-5, 6)$.

Prove that $\triangle RST$ is a right triangle.

[The use of the set of axes on the next page is optional.]

State the coordinates of point $P$ such that quadrilateral $RSTP$ is a rectangle.

**Question 36 is continued on the next page.**

**Question 36 continued**

Prove that your quadrilateral *RSTP* is a rectangle.
[The use of the set of axes below is optional.]

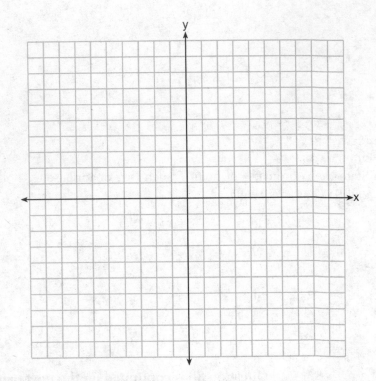

# Answers
# June 2015
## Geometry (Common Core)

---

## Answer Key

---

### PART I

| | | | | | |
|---|---|---|---|---|---|
| **1.** (4) | **5.** (3) | **9.** (1) | **13.** (4) | **17.** (1) | **21.** (4) |
| **2.** (4) | **6.** (2) | **10.** (1) | **14.** (2) | **18.** (1) | **22.** (1) |
| **3.** (3) | **7.** (3) | **11.** (3) | **15.** (3) | **19.** (2) | **23.** (2) |
| **4.** (4) | **8.** (1) | **12.** (4) | **16.** (2) | **20.** (1) | **24.** (3) |

### PART II

**25.** Construct two diameters with the second perpendicular to the first. The endpoints are the vertices of a square.

**26.** m∠MNL = 40°, and ∠MNL and ∠NLO are congruent alternate interior angles so m∠NLO = 40°.

**27.** The coordinates of P are (–2, –3)

**28.** The angle of elevation is 23°

**29.** $\theta = \dfrac{2\pi}{3}$ radians, or 120°

**30.** A reflection is a rigid motion which results in a congruent image, so △ABC ≅ △A'B'C'.

**31.** The height of the flagpole is 6.6 meters.

### PART III

**32.** Alternate interior angles ∠HIG and ∠IGB are congruent.

**33.** △AED ≅ △CEB by the SSS postulate. △AED will map to △CEB under a rotation of 180° about point E.

**34.** The distance between the ranger station and lifeguard is 0.49 miles. Gerald is not correct.

### PART IV

**35.** The volume of the water tower is 7650 ft³. The tower cannot be filled to 85% capacity.

**36.** $\overline{RS} \perp \overline{ST}$, making △RST a right triangle. The coordinates of P are (0, 9).

---

In **PARTS II–IV** you are required to show how you arrived at your answers. For sample methods of solutions, see the *Answers Explained* section.

# Answers Explained

## PART I

1. A cone is generated by rotating a right triangle about one of its legs. As $\overline{ST}$ rotates about point $S$ perpendicular to $\overline{RS}$, it traces out a circle that becomes the base of a cone. $\overline{RT}$ traces out the lateral face of a cone. The height of the cone is $RS$, and point $R$ is the apex.

The correct choice is (**4**).

2. Translations, rotations, and reflections are all rigid motions, which means the pre-image and the image are always congruent. Dilation is a similarity transformation. The image after a dilation will always have the same shape. However, it may have a different size than and not be congruent to the pre-image. Note that you did not need to know the coordinates of $\triangle JKL$ in order to answer this question.

The correct choice is (**4**).

3. The first step is to calculate the length of the radius of the circle. Since a radius is a segment from the center of a circle to any point on the circle, apply the distance formula to calculate the length of radius $QR$ using coordinates $Q(3, -2)$ and $R(7, 1)$.

Let $x_1 = 3$, $x_2 = 7$, $y_1 = -2$, and $y_2 = 1$.

$$
\begin{aligned}
QR &= \sqrt{\left(x_2 - x_1\right)^2 + \left(y_2 - y_1\right)^2} \\
&= \sqrt{(7-3)^2 + (1-(-2))^2} \\
&= \sqrt{16+9} \\
&= \sqrt{25} \\
&= 5
\end{aligned}
$$

The radius of the circle is 5. The diameter of a circle is equal to twice its radius. Since the radius is 5, the diameter is $2 \cdot 5 = 10$.

The correct choice is **(3)**.

4. You can identify each transformation by looking for key features of the preimage and image. When going from figure 1 to figure 2, note the following:

   - The alignment remains unchanged. Both figures have a long vertical axis with $A$ (or $A'$) on top.

   - The orientation doesn't change. The vertices are ordered $A$, $B$, $C$, . . . (or $A'$, $B'$, $C'$, . . .) in a clockwise direction around both figures.

   Rotating figure 1 will change the alignment. So you can rule out rotation. A reflection will always change the orientation. So you can rule out reflection. Of the possible choices, only a translation will preserve the alignment and orientation. The first transformation must be a translation.

   When going from figure 2 to figure 3, observe the following:

   - The slopes of corresponding segments change. The slope of $\overline{A'B'}$ is not equal to the slope of $\overline{A''B''}$.

   - The alignment of the figures changes. The long axis in figure 2 is vertical. In figure 3, it is horizontal.

   - The orientation of the vertices does not change.

   Therefore the transformation cannot be a translation or a reflection. A translation would not change the slope of corresponding segments. A reflection would change the orientation of the vertices. The second transformation must be a rotation.

   The correct choice is **(4)**.

5. The point on the ground to the top of the tree, distance along the ground, and the height of the tree form a right triangle. Trigonometry can be applied to find the height of the tree, $h$.

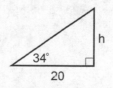

Relative to the 34° angle, the distance along the ground to the tree is the adjacent side and the height ($h$) of the tree is the opposite side. Use the tangent ratio.

$$\tan = \frac{\text{opposite}}{\text{adjacent}}$$

$$\tan(34°) = \frac{h}{20}$$

$$h = 20\tan(34°)$$

$$= 13.4901$$

$$= 13.5$$

The correct choice is **(3)**.

6. A cross-section is the intersection of a plane with a solid figure. The intersection of a plane with a sphere will always be a circle, regardless of the angle at which the plane passes through the sphere. A plane intersecting a circular cone parallel to the cone's base will also produce a circular cross-section. None of the other choices could have a circular cross-section.

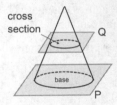

The correct choice is **(2)**.

7. This problem asks for the weight of the contents of a container. We are given the density and dimensions of the contents. The weight can be found using the formula weight = density · volume. Since the material fills the box completely, the volume of the material is equal to the volume of the container. Use the formula for the volume of a rectangular prism to calculate the volume of the container.

$$\text{volume}_{\text{rectangular prism}} = \text{length} \cdot \text{width} \cdot \text{height}$$
$$= 12 \text{ ft} \cdot 8.5 \text{ ft} \cdot 4 \text{ ft}$$
$$= 408 \text{ ft}^3$$
$$\text{weight} = \text{density} \cdot \text{volume}$$
$$= 0.25 \, \frac{\text{lb}}{\text{ft}^3} \cdot 408 \text{ ft}^3$$
$$= 102 \text{ lb}$$

The correct choice is **(3)**.

8. The two triangles $\triangle CEG$ and $\triangle FDG$ are similar. $\angle C$ and $\angle F$ both intercept $\overset{\frown}{ED}$. Since inscribed angles that intercept the same arc are congruent, you know $\angle C \cong \angle F$. Since $\angle CGE$ and $\angle FGD$ are vertical angles, they are congruent. Therefore $\triangle CEG \sim \triangle FDG$ by the AA postulate. All corresponding angles are congruent, and all pairs of corresponding sides are proportional. Now examine the choices.

- Choice (1): $\overline{CG}$ and $\overline{FG}$ are corresponding sides but are not necessarily congruent because the triangles may not be congruent.

- Choice (2): $\angle CEG$ and $\angle FDG$ are corresponding angles. So they are congruent.

- Choice (3): $CE$ and $FD$ are corresponding lengths, as are $EG$ and $DG$. Therefore their ratios must be equal.

- Choice (4): $\triangle CEG \sim \triangle FDG$ by the AA postulate.

The correct choice is **(1)**.

9. Perpendicular lines have negative reciprocal slopes. So the first step is to identify the slope of the given line by rewriting it in the slope-intercept form $y = mx + b$.

$$2x - y = 7$$
$$-y = -2x + 7$$
$$y = 2x - 7$$

The slope of the given line is 2. The negative reciprocal of 2 is $-\frac{1}{2}$. So a line perpendicular to the given line will have a slope of $-\frac{1}{2}$. Choice (1), $y = -\frac{1}{2}x + 6$, is the only choice with a slope of $-\frac{1}{2}$.

The correct choice is **(1)**.

10. A regular polygon will rotate onto itself when the rotation angle is any multiple of its central angle. The smallest possible rotation will be equal to the central angle. The central angle for each figure is calculated using the formula:

$$\text{central angle} = \frac{360°}{\text{number of sides}}$$

| Figure | Number of Sides | Central Angle |
|--------|-----------------|---------------|
| octagon | 8 | 45° |
| decagon | 10 | 36° |
| hexagon | 6 | 60° |
| pentagon | 5 | 72° |

The octagon has a 45° central angle, which is shown in the accompanying figure. After the 45° rotation, vertex $A$ will map to $B$, vertex $B$ will map to $C$, and so on.

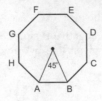

The correct choice is **(1)**.

**11.** A segment that intersects two sides of a triangle and is parallel to the third side will form a smaller triangle similar to the original, so $\triangle AEB \sim \triangle ADC$. To identify the corresponding parts more easily, sketch the two triangles separately. The length $AD$ is found by adding $AE + ED$, $9 + 5 = 14$.

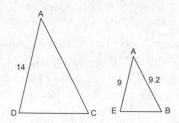

Corresponding parts of similar triangles are proportional. You can form a proportion involving the unknown side $\overline{AC}$:

$$\frac{AD}{AE} = \frac{AC}{AB}$$

$$\frac{14}{9} = \frac{AC}{9.2}$$

$9 \cdot AC = 14 \cdot 9.2$     cross-multiply

$9 \cdot AC = 128.8$

$AC - 14.311$

$AC = 14.3$     round to the nearest tenth

The correct choice is **(3)**.

**12.** The definitions for the sine and cosine are as follows:

$$\sin = \frac{\text{opposite}}{\text{hypotenuse}} \qquad \cos = \frac{\text{adjacent}}{\text{hypotenuse}}$$

Using the accompanying figure, write out each of the ratios in terms of the side lengths of the triangle.

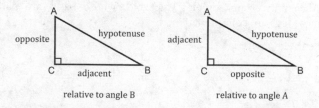

relative to angle B                  relative to angle A

$$\sin A = \frac{BC}{AB} \qquad\qquad \sin B = \frac{AC}{AB}$$

$$\cos A = \frac{AC}{AB} \qquad\qquad \cos B = \frac{BC}{AB}$$

By comparing each of the choices, you can see that $\sin A = \cos B = \dfrac{BC}{AB}$. You could also have applied the cofunction relationship. It states that given two complementary angles, the sine of one equals the cosine of the other.

The correct choice is (**4**).

13. Use the accompanying figure of quadrilateral $ABCD$ to compare each of the choices to the properties that are sufficient to prove a quadrilateral is a parallelogram.

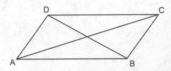

- Choice (1): $\overline{AC}$ and $\overline{BD}$ bisect each other tells us that the diagonals bisect each other. This is one of the properties sufficient to prove that a quadrilateral is a parallelogram.

- Choice (2): $\overline{AB} \cong \overline{CD}$ and $\overline{BC} \cong \overline{AD}$ tells us that the opposite sides are congruent. This is one of the properties sufficient to prove that the figure is a parallelogram.

- Choice (3): $\overline{AB} \cong \overline{CD}$ and $\overline{AB} \parallel \overline{CD}$ indicates that one pair of opposite sides are both parallel and congruent. This is sufficient to prove that the figure is a parallelogram.

- Choice (4): $\overline{AB} \cong \overline{CD}$ and $\overline{BC} \parallel \overline{AD}$ tells us that one pair of opposite sides is congruent, and that the other pair of sides is parallel. This is not sufficient to prove that a quadrilateral is a parallelogram. A counterexample is an isosceles trapezoid, whose bases are parallel and whose legs are congruent.

The correct choice is (**4**).

**14.** The equation can be rewritten in center–radius form by completing the square with the $y$-terms. The coefficient of the $y$-term is 6. A constant of $\left(\frac{1}{2} \cdot 6\right)^2$ , or 9, is needed to complete the square. First add 9 to both sides of the equation. Then factor the $y$-terms.

$$x^2 + y^2 + 6y = 7$$
$$x^2 + y^2 + 6y + 9 = 7 + 9$$
$$x^2 + (y + 3)^2 = 16$$

The center–radius form of the equation of a circle is $(x - h)^2 + (y - k)^2 = r^2$, where $(h, k)$ are the coordinates of the center of the circle and $r$ is the radius. Since the equation of the circle is in center–radius form, you can see that $h = 0$, $k = -3$, and $r^2 = 16$. The center has coordinates $(0, -3)$, and the radius is $\sqrt{16}$, or 4.

The correct choice is **(2)**.

**15.** The two labeled triangles are shown below. They have two pairs of corresponding sides that are in the same ratio.

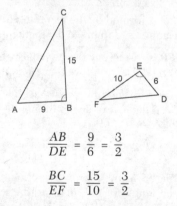

$$\frac{AB}{DE} = \frac{9}{6} = \frac{3}{2}$$

$$\frac{BC}{EF} = \frac{15}{10} = \frac{3}{2}$$

The included angles, $\angle C$ and $\angle E$, are given as congruent. The triangles are therefore similar by the SAS similarity postulate. A correct similarity statement is $\triangle ABC \sim \triangle DEF$. Note that the pairs of corresponding sides must come in the same order when naming each triangle. $\overline{AB}$ corresponds to $\overline{DE}$, and $\overline{CB}$ corresponds to $\overline{FE}$. Choice (1) cannot be proven. Choices (2) and (4) use ratios of sides that are *not* corresponding.

The correct choice is **(3)**.

**16.** When a figure is dilated, all lengths are multiplied by the scale factor but the angle measures are unchanged. Choices (3) and (4) can therefore be eliminated. Since $\triangle ABC$ is dilated by a scale factor of 3, all lengths in $\triangle A'B'C'$ are 3 times greater than those in $\triangle ABC$:

$$A'B' = 3AB$$

$$B'C' = 3BC$$

$$C'A' = 3CA$$

The correct choice is (**2**).

**17.** Two segments intersected by a transversal can be proven to be parallel if alternate interior angles are congruent, corresponding angles are congruent, or same side interior angles are supplementary. In the given figure, there are two transversals, $\overline{BF}$ and $\overline{CF}$. Compare each choice with the conditions that prove two segments are parallel. You need to find two angles formed by only one transversal.

- Choice (1): $\angle CFG$ and $\angle FCB$ are congruent alternate interior angles formed by transversal $\overline{CF}$. This condition proves $\overline{ABCD} \,/\!/\, \overline{EFG}$.

- Choice (2): $\angle BFC$ is formed by the two transversals. So it is not one of the angles that can be used to prove the segments are parallel.

- Choice (3): $\angle CFB$ is formed by the two transversals. So it is not one of the angles that can be used to prove the segments are parallel.

- Choice (4): $\angle CBF$ is formed by transversal $\overline{BF}$, and $\angle GFC$ is formed by transversal $\overline{CF}$. This pair of angles cannot be used to prove the segments are parallel.

The correct choice is (**1**).

**18.** In a dilation, all lengths in the preimage are multiplied by the scale factor. The ratio of the length of the image to the length of the preimage, $\dfrac{CD}{AB}$, is equal to the scale factor.

The distance from any point in the preimage to the center of dilation is also multiplied by the scale factor. So the ratio of the distances to point $E$ is also equal to the scale factor. The ratios of corresponding distances to the center are $\dfrac{EC}{EA}$ and $\dfrac{ED}{EB}$.

The correct choice is (**1**).

**19.** A cube has 6 congruent square faces. Each face of the cube has a length and a width of 12 feet.

First find the surface area of one face. Then find the total surface area of the cube.

$$\text{surface area}_{\text{face}} = s^2$$

$$= (12 \text{ ft})^2$$

$$= 144 \text{ ft}^2$$

$$\text{surface area}_{\text{total}} = 6 \cdot \text{surface area}_{\text{face}}$$

$$= 6 \cdot 144 \text{ ft}^2$$

$$= 864 \text{ ft}^2$$

Now divide the total surface area of the cube by the number of square feet that 1 gallon of paint will cover.

$$\text{number of gallons} = \frac{\text{surface area}_{\text{total}}}{450 \text{ ft}^2/\text{gallon}}$$

$$= \frac{864 \text{ ft}^2}{450 \text{ ft}^2/\text{gallon}}$$

$$= 1.92 \text{ gallons}$$

This number must be rounded up to 2 gallons to determine the number of gallons of paint that must be purchased.

The correct choice is **(2)**.

**20.** The measure of each angle can be stated in terms of the measure of an intercepted arc. Keep in mind the following relationships.

First, the measure of an inscribed angle equals half the measure of the intercepted arc.

Second, the measure of an angle formed by a tangent and chord to that tangent equals half the measure of the intercepted arc.

Third, a diameter intercepts a 180° arc.

- Choice (1): $m\angle ACB = \frac{1}{2}m\widehat{AB}$ and $m\angle BCD = \frac{1}{2}m\widehat{BC}$. Not enough information is provided to conclude that the two arcs are congruent. So you cannot conclude that the angles are congruent.

- Choice (2): $m\angle ABC = \frac{1}{2}m\widehat{AEC}$ and $m\angle ACD = \frac{1}{2}m\widehat{ABC}$. The two arcs both measure 180° because they are also intercepted by diameter $\overline{AOC}$. Therefore the angles have the same measure and $\angle ABC \cong \angle ACD$.

- Choice (3): $m\angle BAC = \frac{1}{2}m\widehat{BC}$ and $m\angle DCB = \frac{1}{2}m\widehat{BC}$. Therefore the angles have the same measure and $\angle BAC \cong \angle DCB$.

- Choice (4): $m\angle CBA = \frac{1}{2}m\widehat{AEC}$ and $m\angle AEC = \frac{1}{2}m\widehat{ABC}$. The two arcs both measure 180° because they are intercepted by diameter $\overline{AOC}$. Therefore the angles have the same measure and $\angle CBA \cong \angle AEC$.

The correct choice is **(1)**.

**21.** All corresponding lengths of similar triangles are proportional as are their perimeters. First find the scale factor using a pair of proportional sides. Then apply the scale factor to determine the perimeter of $\triangle DEC$.

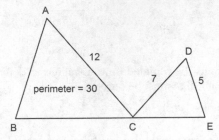

From the similarity statement $\triangle ABC \sim \triangle DEC$, $\overline{AC}$ and $\overline{DC}$ are corresponding sides.

$$\text{scale factor} = \frac{DC}{AC}$$

$$= \frac{7}{12}$$

$$= 0.583$$

perimeter $\triangle DCE$ = perimeter $\triangle ABC \cdot$ scale factor

$$= 30(0.583)$$

$$= 17.49$$

$$= 17.5$$

The correct choice is **(4)**.

22. A dilation about the origin will preserve the slope of a line and will multiply the $y$-intercept of the line by the scale factor. First rewrite the equation in slope-intercept form.

$$3y = -2x + 8$$

$$y = -\frac{2}{3}x + \frac{8}{3}$$

The slope of the dilated line must also be $-\frac{2}{3}$. Rewrite each of the choices in slope-intercept form to determine its slope.

- Choice (1):

$$2x + 3y = 5$$

$$3y = -2x + 5$$

$$y = -\frac{2}{3}x + \frac{5}{3}$$

$$\text{slope} = -\frac{2}{3}$$

- Choice (2):

$$2x - 3y = 5$$

$$-3y = -2x + 5$$

$$y = \frac{2}{3}x - \frac{5}{3}$$

$$\text{slope} = \frac{2}{3}$$

- Choice (3):

$$3x + 2y = 5$$

$$2y = -3x + 5$$

$$y = -\frac{3}{2}x + \frac{5}{2}$$

$$\text{slope} = -\frac{3}{2}$$

- Choice (4):

$$3x - 2y = 5$$

$$-2y = -3x + 5$$

$$y = \frac{3}{2}x - \frac{5}{2}$$

$$\text{slope} = \frac{3}{2}$$

The line with a slope of $-\frac{2}{3}$ is choice (1).

The correct choice is (**1**).

23. The sum of the short arcs of each sector is equal to the circumference of the circle. The dimension labeled $x$ in the final figure is approximately $\frac{1}{2}$ the circumference of the circle.

$$x = \frac{1}{2}\text{ circumference}$$

$$= \frac{1}{2}(2\pi r) \qquad \text{substitute } 2\pi r \text{ for the circumference}$$

$$= \frac{1}{2}(2\pi)(5)$$

$$= 15.707$$

$$= 16$$

The correct choice is (**2**).

**24.** Two triangles can be proven congruent by the congruence postulates (SSS, SAS, ASA, AAS, and HL) or by showing that there is a sequence of rigid motions that will map one triangle onto another. Examine each choice.

- Choice (1): Two pairs of congruent sides is not sufficient to demonstrate one of the congruence postulates.

- Choice (2): Three pairs of congruent angles would indicate similar, but not necessarily congruent, triangles.

- Choice (3): This sequence of rigid motions maps each side of $\triangle ABC$ to a side of $\triangle DEF$. Therefore all three pairs of corresponding sides are congruent. The triangles are congruent by SSS.

- Choice (4): Note that points $A$ and $D$ are mapped to each other, not $\angle A$ and $\angle D$. This sequence of rigid motions maps only one pair of corresponding angles and one pair of corresponding sides. One angle and one side are not sufficient to demonstrate any of the congruence postulates.

The correct choice is **(3)**.

## PART II

**25.** Begin by constructing a diameter through point *T*.

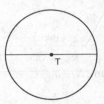

Next construct the perpendicular bisector of the diameter, which will also be a diameter.

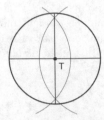

The two diameters are the diagonals of the square. The intersections of the diameters with the circle are the vertices of the square.

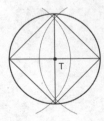

**26.** When asked to explain why an angle or a segment has a certain measure, be sure to give a reason in words for any of the geometric relationships that are required.

Consecutive angles of a parallelogram are supplementary.

$$m\angle M + m\angle MNL + m\angle ONL = 180°$$
$$118° + m\angle MNL + 22° = 180°$$
$$140° + m\angle MNL = 180°$$
$$m\angle MNL = 40°$$

$\angle MNL$ and $\angle NLO$ are congruent alternate interior angles formed by transversal $\overline{LN}$ and parallel sides $\overline{LM}$ and $\overline{NO}$.

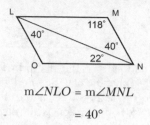

$$m\angle NLO = m\angle MNL$$

$$= 40°$$

**27.** Find the $x$- and $y$-coordinates of point $P$ using the formula

$$\text{ratio} = \frac{x - x_1}{x_2 - x} \qquad \text{ratio} = \frac{y - y_1}{y_2 - y}$$

Using $A(-6, -5)$ and $B(4, 0)$, $x_1 = -6$, $x_2 = 4$, $y_1 = -5$, and $y_2 = 0$. The coordinates of the point are $P(x, y)$. The ratio is $\frac{2}{3}$.

$$\text{ratio} = \frac{x - x_1}{x_2 - x} \qquad\qquad \text{ratio} = \frac{y - y_1}{y_2 - y}$$

$$\frac{2}{3} = \frac{x - (-6)}{4 - x} \qquad\qquad \frac{2}{3} = \frac{y - (-5)}{0 - y}$$

$$3(x + 6) = 2(4 - x) \qquad\qquad 3(y + 5) = 2(0 - y)$$

$$3x + 18 = 8 - 2x \qquad\qquad 3y + 15 = -2y$$

$$5x + 18 = 8 \qquad\qquad 5y + 15 = 0$$

$$5x = -10 \qquad\qquad 5y = -15$$

$$x = -2 \qquad\qquad y = -3$$

The coordinates of point $P$ are **(–2, –3)**.

**28.** The angle of elevation formed by the ramp and the ground is indicated by $x$ in the following drawing.

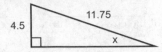

The ground, ramp, and side of the platform form a right triangle. An inverse trigonometric function can be used to calculate the angle of elevation. Relative to the angle of elevation the 4.5 ft dimension is the opposite and the 11.75 ft dimension is the hypotenuse. The sine ratio applies here.

$$\sin(x) = \frac{4.5}{11.75}$$

$$x = \sin^{-1}\left(\frac{4.5}{11.75}\right)$$

$$= 22.51°$$

$$= 23°$$

To the *nearest degree*, the angle of elevation is **23°**.

**29.** The area of a sector can be calculated in two ways. The first uses the formula $A = \frac{1}{2}r^2\theta$, where $r$ is the radius of the circle and $\theta$ is the measure of the central angle in radians.

$$A = \frac{1}{2}r^2\theta$$

$$12\pi = \frac{1}{2}(6^2)\theta$$

$$12\pi = 18\theta$$

$$\theta = \frac{12\pi}{18} \text{ radians}$$

$$\theta = \frac{2\pi}{3} \text{ radians}$$

The second way to calculate the area of a sector uses the formula $A = \left(\dfrac{n}{360}\right)\pi r^2$, where $n$ is the measure of the central angle in degrees and $n$ is the radius of the circle.

$$A = \left(\frac{n}{360}\right)\pi r^2$$

$$12\pi = \left(\frac{n}{360}\right)\pi(6^2)$$

$$12\pi = \left(\frac{n}{360}\right)36\pi$$

$$12\pi = \frac{n \cdot \pi}{10}\pi r^2$$

$$n = 120 \text{ degrees}$$

$m\angle AOC$ is $\theta = \dfrac{2\pi}{3}$ **radians** or **120°**.

30. A reflection is a rigid motion that preserves length, so $\overline{AB} \cong \overline{A'B'}$, $\overline{BC} \cong \overline{B'C'}$, and $\overline{CA} \cong \overline{C'A'}$. Three pairs of corresponding sides are congruent, so $\triangle ABC \cong \triangle A'B'C'$ by the SSS postulate.

31. The flagpole and Tim both stand vertically upright, making right angles with the ground. The sun's rays strike both the flagpole and Tim at the same angle, forming a pair of triangles as shown in the figure. The height of the flagpole is indicated by $x$.

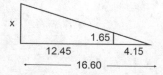

The distance from Tim to the tip of his shadow can be calculated from the given information.

$$\text{distance} = 16.60 \text{ m} - 12.45 \text{ m}$$

$$= 4.15 \text{ m}$$

The two triangles are similar by the AA postulate—one pair of congruent right angles and a shared angle at the right. The height of the flagpole is found using a proportion of corresponding parts.

$$\frac{x}{16.60} = \frac{1.65}{4.15}$$

$$4.15x = 16.60(1.65)$$

$$4.15x = 27.39$$

$$x = 6.6$$

The height of the flagpole is **6.6 meters**.

Another approach would be to use the inverse tangent with the small triangle to calculate the angle of elevation. Then use a tangent ratio with the large triangle to calculate the height of the flagpole.

# PART III

**32.** To explain why $\overline{AB} \parallel \overline{CD}$, you need to demonstrate that alternate interior angles are congruent, that corresponding angles are congruent, or that same side interior angles are supplementary. Any one of these can be demonstrated. Alternate interior angles will be demonstrated here. Remember that a complete and valid explanation must not presume $\overline{AB} \parallel \overline{CD}$ and that a written reason for each geometric relationship must be included.

$$m\angle HIG + m\angle DIG = 180°$$    linear pairs of angles are supplementary

$$m\angle HIG + 115° = 180°$$

$$m\angle HIG = 65°$$

$$m\angle HGI = m\angle HIG$$    base angles of an isosceles triangle are congruent

$$= 65°$$

$$m\angle IGB + m\angle HGI + m\angle EGB = 180°$$    sum of angles around a straight line equals 180°

$$m\angle IGB + 65° + 50° = 180°$$

$$m\angle IGB + 115° = 180°$$

$$m\angle IGB = 65°$$

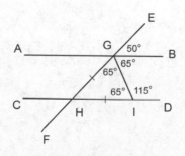

$\overline{AB} \parallel \overline{CD}$ because alternate interior angles $\angle HIG$ and $\angle IGB$ are congruent.

**33.** The parallelogram properties that will be used are opposite sides are congruent and diagonals bisect each other. Three pairs of corresponding sides can be shown congruent, which leads to congruence by SSS.

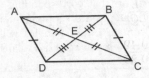

| Statement | Reason |
|---|---|
| 1. Parallelogram $ABCD$ with diagonals $\overline{AC}$ and $\overline{BD}$ intersecting at $E$ | 1. Given |
| 2. $\overline{AD} \cong \overline{BC}$ | 2. Opposite sides of a parallelogram are congruent |
| 3. $\overline{AE} \cong \overline{EC}$ and $\overline{DE} \cong \overline{EB}$ | 3. Diagonals of a parallelogram bisect each other |
| 4. $\triangle AED \cong \triangle CEB$ | 4. SSS |

$\triangle AED$ will map to $\triangle CEB$ under a rotation of $180°$ about point $E$.

**34.** The distance from the park ranger station to the lifeguard chair, $PL$, can be calculated using the Pythagorean theorem.

$$PL^2 + LC^2 = PC^2$$

$$PL^2 + 0.25^2 = 0.55^2$$

$$PL^2 + 0.0625 = 0.3025$$

$$PL^2 = 0.24$$

$$PL = \sqrt{0.24}$$

$$= 0.489$$

$$= 0.49 \text{ mi}$$

To find the distance from the first aid station to the campground, $FC$, use the fact that the altitude to the hypotenuse of a right triangle is the mean proportional to the parts of the hypotenuse.

$$\frac{FL}{PL} = \frac{PL}{LC}$$

$$\frac{FL}{0.49} = \frac{0.49}{0.25}$$

$0.25FL = (0.49)(0.49)$     cross multiply

$0.25FL = 0.24$

$FL = 0.96$

$FC = FL + LC$     $FC$ equals the sum of its parts

$\quad = 0.96 + 0.25$

$\quad = 1.21$

To the *nearest hundredth of a mile*, the distance between the park ranger station and the lifeguard chair is **0.49 miles**.

Gerald is not correct. The distance from the first aid station to the campground is only 1.21 miles.

## PART IV

**35.** The volume of the water tower is the sum of the volumes of the three parts—cone, cylinder, and hemisphere. Calculate the volume of each part separately using the formulas found on the reference sheet.

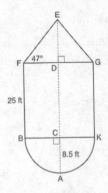

Cone: The radius of the cone is equal to the radius of the hemisphere, or 8.5 ft. The height of the cone must be calculated using the tangent ratio.

$$\tan = \frac{\text{opposite}}{\text{adjacent}}$$

$$\tan(47°) = \frac{h}{8.5}$$

$$h = 8.5 \cdot \tan(47°)$$

$$h = 9.1151$$

$$V_{cone} = \frac{1}{3}\pi r^2 h$$

$$= \frac{1}{3}\pi(8.5 \text{ ft})^2(9.1151 \text{ ft})$$

$$= 689.6486 \text{ ft}^3$$

Cylinder: The radius of the cylinder is 8.5 ft, and its height is 25 ft.

$$V_{cylinder} = \pi r^2 h$$
$$= \pi (8.5 \text{ ft})^2 (25 \text{ ft})$$
$$= 5674.5017 \text{ ft}^3$$

Hemisphere: The volume of the hemisphere is $\dfrac{1}{2}$ the volume of the sphere whose radius is 8.5 ft.

$$V_{hemisphere} = \left(\frac{1}{2}\right)\left(\frac{4}{3}\right)\pi r^3$$
$$= \left(\frac{1}{2}\right)\left(\frac{4}{3}\right)\pi (8.5 \text{ ft})^3$$
$$= 1286.2203 \text{ ft}^3$$

The volume of the water tower equals the sum of the volume of its parts.

$$V_{water\ tower} = V_{cone} + V_{cylinder} + V_{hemisphere}$$
$$= 689.6486 \text{ ft}^3 + 5674.5017 \text{ ft}^3 + 1286.2203 \text{ ft}^3$$
$$= 7650.3706 \text{ ft}^3$$
$$= 7650 \text{ ft}^3$$

To calculate the weight of the water in the tank, use the relationship weight = volume · density. Remember to multiply the volume by 0.85 to account for the fact that you want to fill the water tower only 85% full.

$$weight = (0.85)(volume) \cdot density$$
$$= (0.85)(7650 \text{ ft}^3)\left(62.4 \ \frac{lb}{ft^3}\right)$$
$$= 405{,}756 \text{ lb}$$

To the *nearest cubic foot*, the volume of the water tower is **7650 ft³**.

The water tower cannot be filled to 85% capacity because that would exceed the weight limit of 400,000 lb.

**36.** Prove that $\triangle RST$ is a right triangle by using the slope formula and the coordinates $R(6, -1)$, $S(1, -4)$, and $T(-5, 6)$ to show that $\overline{ST}$ is perpendicular to $\overline{RS}$.

$$\text{slope} = \frac{y_2 - y_1}{x_2 - x_1}$$

$$\text{slope } \overline{ST} = \frac{6 - (-4)}{-5 - 1}$$

$$= \frac{10}{-6}$$

$$= -\frac{5}{3}$$

$$\text{slope } \overline{RS} = \frac{-4 - (-1)}{1 - 6}$$

$$= \frac{-3}{-5}$$

$$= \frac{3}{5}$$

$\overline{RS}$ and $\overline{ST}$ have negative reciprocal slopes, making them perpendicular. So $\triangle RST$ is a right triangle.

The easiest way to locate point $P$ to show that $RSTP$ is a rectangle is graphically. Opposite sides of a parallelogram are parallel and congruent. So travel along the graph from point $R$ the same distance and slope as $\overline{ST}$, which is 10 units up and 6 units left. $\overline{RP}$ will be parallel and congruent to $\overline{ST}$. Since you already know the figure has a right angle, $RSTP$ must be a rectangle.

$P$ has the coordinates $(6 - 6, -1 + 10)$ or $(0, 9)$.

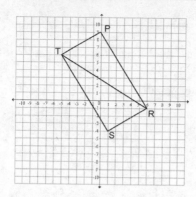

As an alternative, $RSTP$ can be proven to be a rectangle by algebraically showing the slopes and lengths of $\overline{ST}$ and $\overline{RP}$ are equal, making then parallel and congruent:

$$\text{slope } \overline{RP} = \frac{9-(-1)}{0-6}$$

$$= \frac{10}{-6}$$

$$= -\frac{5}{3}$$

Lengths are calculated using the distance formula $\sqrt{(x_1-x_2)^2+(y_1-y_2)^2}$.

$$\text{Length } \overline{RP} = \sqrt{(0-6)^2+(9-(-1))^2}$$

$$= \sqrt{36+100}$$

$$= \sqrt{136}$$

$$\text{Length } \overline{ST} = \sqrt{(-5-1)^2+(6-(-4))^2}$$

$$= \sqrt{36+100}$$

$$= \sqrt{136}$$

$\overline{ST}$ and $\overline{RP}$ are parallel because their slopes are both $-\dfrac{5}{3}$. They are congruent because their lengths are both $\sqrt{136}$. $RSTP$ also has a right angle at $S$. Therefore $RSTP$ is a rectangle.

The coordinates of point $P$ are **(0, 9)**.

| Topic | Question Numbers | Number of Points | Your Points | Your Percentage |
|---|---|---|---|---|
| 1. Basic Angle and Segment Relationships | 17, 32 | 2 + 4 = 6 | | |
| 2. Angle and Segment Relationships in Triangles and Polygons | 34 | 4 | | |
| 3. Constructions | 25 | 2 | | |
| 4. Transformations | 2, 4, 10, 16, 18, 22, 33 | 2 + 2 + 2 + 2 + 2 + 2 + 4 = 16 | | |
| 5. Triangle Congruence | 24, 30 | 2 + 2 = 4 | | |
| 6. Line, Segments and Circles on the Coordinate Plane | 3, 9, 14, 27 | 2 + 2 + 2 + 2 = 8 | | |
| 7. Similarity | 11, 15, 21, 31, 34 | 2 + 2 + 2 + 2 + 4 = 12 | | |
| 8. Trigonometry | 5, 12, 28, 31 | 2 + 2 + 2 + 2 = 8 | | |
| 9. Parallelograms | 13, 26, 33 | 2 + 2 + 4 = 8 | | |
| 10. Coordinate Geometry Proofs | 36 | 6 | | |
| 11. Volume | 1, 6, 23 | 2 + 2 + 2 = 6 | | |
| 12. Modeling | 7, 19, 35 | 2 + 2 + 6 = 10 | | |

# HOW TO CONVERT YOUR RAW SCORE TO YOUR GEOMETRY REGENTS EXAMINATION SCORE

The conversion chart below can be used to determine your final score on the June 2015 Regents Exam in Geometry. To find your final exam score, locate in the "Raw Score" column the total number of points you scored out of a possible 86. Then locate in the adjacent column to the right the scale score that corresponds to your raw score. The scale score is your final score.

| Raw Score | Scale Score | Performance Level | Raw Score | Scale Score | Performance Level | Raw Score | Scale Score | Performance Level |
|---|---|---|---|---|---|---|---|---|
| 86 | 100 | 5 | 57 | 79 | 3 | 28 | 60 | 2 |
| 85 | 99 | 5 | 56 | 79 | 3 | 27 | 59 | 2 |
| 84 | 98 | 5 | 55 | 78 | 3 | 26 | 58 | 2 |
| 83 | 97 | 5 | 54 | 78 | 3 | 25 | 56 | 2 |
| 82 | 96 | 5 | 53 | 77 | 3 | 24 | 55 | 2 |
| 81 | 95 | 5 | 52 | 77 | 3 | 23 | 54 | 1 |
| 80 | 94 | 5 | 51 | 77 | 3 | 22 | 52 | 1 |
| 79 | 93 | 5 | 50 | 76 | 3 | 21 | 51 | 1 |
| 78 | 92 | 5 | 49 | 76 | 3 | 20 | 49 | 1 |
| 77 | 91 | 5 | 48 | 75 | 3 | 19 | 47 | 1 |
| 76 | 90 | 5 | 47 | 75 | 3 | 18 | 46 | 1 |
| 75 | 90 | 5 | 46 | 74 | 3 | 17 | 44 | 1 |
| 74 | 89 | 5 | 45 | 73 | 3 | 16 | 42 | 1 |
| 73 | 88 | 5 | 44 | 73 | 3 | 15 | 40 | 1 |
| 72 | 88 | 5 | 43 | 72 | 3 | 14 | 38 | 1 |
| 71 | 87 | 5 | 42 | 72 | 3 | 13 | 36 | 1 |
| 70 | 86 | 5 | 41 | 71 | 3 | 12 | 34 | 1 |
| 69 | 86 | 5 | 40 | 70 | 3 | 11 | 32 | 1 |
| 68 | 85 | 5 | 39 | 70 | 3 | 10 | 29 | 1 |
| 67 | 84 | 4 | 38 | 69 | 3 | 9 | 27 | 1 |
| 66 | 84 | 4 | 37 | 68 | 3 | 8 | 24 | 1 |
| 65 | 83 | 4 | 36 | 68 | 3 | 7 | 22 | 1 |
| 64 | 83 | 4 | 35 | 67 | 3 | 6 | 19 | 1 |
| 63 | 82 | 4 | 34 | 66 | 3 | 5 | 16 | 1 |
| 62 | 82 | 4 | 33 | 65 | 3 | 4 | 13 | 1 |
| 61 | 81 | 4 | 32 | 64 | 2 | 3 | 10 | 1 |
| 60 | 81 | 4 | 31 | 63 | 2 | 2 | 7 | 1 |
| 59 | 80 | 4 | 30 | 62 | 2 | 1 | 3 | 1 |
| 58 | 80 | 4 | 29 | 61 | 2 | 0 | 0 | 1 |

# Examination August 2015
## Geometry (Common Core)

## GEOMETRY REFERENCE SHEET

| | |
|---|---|
| 1 inch = 2.54 centimeters | 1 ton = 2000 pounds |
| 1 meter = 39.37 inches | 1 cup = 8 fluid ounces |
| 1 mile = 5280 feet | 1 pint = 2 cups |
| 1 mile = 1760 yards | 1 quart = 2 pints |
| 1 mile = 1.609 kilometers | 1 gallon = 4 quarts |
| 1 kilometer = 0.62 mile | 1 gallon = 3.785 liters |
| 1 pound = 16 ounces | 1 liter = 0.264 gallon |
| 1 pound = 0.454 kilogram | 1 liter = 1000 cubic centimeters |
| 1 kilogram = 2.2 pounds | |

| | |
|---|---|
| Triangle | $A = \frac{1}{2}bh$ |
| Parallelogram | $A = bh$ |
| Circle | $A = \pi r^2$ |
| Circle | $C = \pi d$ or $C = 2\pi r$ |
| General Prisms | $V = Bh$ |
| Cylinder | $V = \pi r^2 h$ |
| Sphere | $V = \frac{4}{3}\pi r^3$ |

| Cone | $V = \frac{1}{3}\pi r^2 h$ |
|---|---|
| Pyramid | $V = \frac{1}{3}Bh$ |
| Pythagorean Theorem | $a^2 + b^2 = c^2$ |
| Quadratic Formula | $x = \dfrac{-b \pm \sqrt{b^2 - 4ac}}{2a}$ |
| Arithmetic Sequence | $a_n = a_1 + (n-1)d$ |
| Geometric Sequence | $a_n = a_1 r^{n-1}$ |
| Geometric Series | $S_n = \dfrac{a_1 - a_1 r^n}{1-r}$ where $r \neq 1$ |
| Radians | 1 radian = $\dfrac{180}{\pi}$ degrees |
| Degrees | 1 degree = $\dfrac{\pi}{180}$ radians |
| Exponential Growth/Decay | $A = A_0 e^{k(t-t_0)} + B_0$ |

## PART I

Answer all 24 questions in this part. Each correct answer will receive 2 credits. No partial credit will be allowed. For each statement or question, write in the space provided the numeral preceding the word or expression that best completes the statement or answers the question. [48 credits]

1 A parallelogram must be a rectangle when its

(1) diagonals are perpendicular
(2) diagonals are congruent
(3) opposite sides are parallel
(4) opposite sides are congruent

1 _____

2 If $\triangle A'B'C'$ is the image of $\triangle ABC$, under which transformation will the triangles *not* be congruent?

(1) reflection over the $x$-axis
(2) translation to the left 5 and down 4
(3) dilation centered at the origin with scale factor 2
(4) rotation of 270° counterclockwise about the origin

2 _____

3 If the rectangle below is continuously rotated about side $w$, which solid figure is formed?

(1) pyramid
(2) rectangular prism
(3) cone
(4) cylinder

3 _____

4 Which expression is always equivalent to sin $x$ when
$0° < x < 90°$?

(1) $\cos(90° - x)$       (3) $\cos(2x)$
(2) $\cos(45° - x)$       (4) $\cos x$          4 _____

5 In the diagram below, a square is graphed in the
coordinate plane.

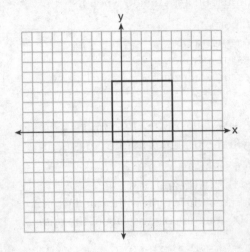

A reflection over which line does *not* carry the square
onto itself?

(1) $x = 5$       (3) $y = x$
(2) $y = 2$       (4) $x + y = 4$         5 _____

**6** The image of △*ABC* after a dilation of scale factor *k* centered at point *A* is △*ADE*, as shown in the diagram below.

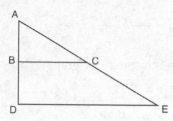

Which statement is always true?

(1) $2AB = AD$            (3) $AC = CE$

(2) $\overline{AD} \perp \overline{DE}$            (4) $\overline{BC} \parallel \overline{DE}$          6 _____

**7** A sequence of transformations maps rectangle *ABCD* onto rectangle *A″B″C″D″*, as shown in the diagram below.

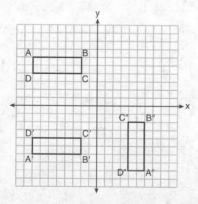

Which sequence of transformations maps *ABCD* onto *A′B′C′D′* and then maps *A′B′C′D′* onto *A″B″C″D″*?

(1) a reflection followed by a rotation

(2) a reflection followed by a translation

(3) a translation followed by a rotation

(4) a translation followed by a reflection          7 _____

**8** In the diagram of parallelogram *FRED* shown below, $\overline{ED}$ is extended to *A*, and $\overline{AF}$ is drawn such that $\overline{AF} \cong \overline{DF}$.

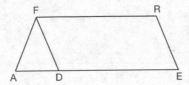

If m∠R = 124°, what is m∠AFD?

(1) 124°  (3) 68°
(2) 112°  (4) 56°

8 _____

**9** If $x^2 + 4x + y^2 - 6y - 12 = 0$ is the equation of a circle, the length of the radius is

(1) 25  (3) 5
(2) 16  (4) 4

9 _____

**10** Given $\overline{MN}$ shown below, with $M(-6, 1)$ and $N(3, -5)$, what is an equation of the line that passes through point $P(6, 1)$ and is parallel to $\overline{MN}$?

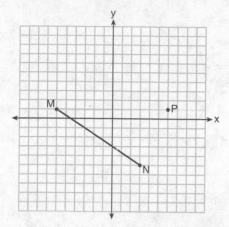

(1) $y = -\dfrac{2}{3}x + 5$        (3) $y = \dfrac{3}{2}x + 7$

(2) $y = -\dfrac{2}{3}x - 3$        (4) $y = \dfrac{3}{2}x - 8$      10 _____

**11** Linda is designing a circular piece of stained glass with a diameter of 7 inches. She is going to sketch a square inside the circular region.

To the *nearest tenth of an inch*, the largest possible length of a side of the square is

(1) 3.5        (3) 5.0

(2) 4.9        (4) 6.9      11 _____

**12** In the diagram shown below, $\overline{AC}$ is tangent to circle $O$ at $A$ and to circle $P$ at $C$, $\overline{OP}$ intersects $\overline{AC}$ at $B$, $OA = 4$, $AB = 5$, and $PC = 10$.

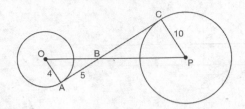

What is the length of $\overline{BC}$?

(1) 6.4      (3) 12.5

(2) 8      (4) 16      12 _____

**13** In the diagram below, which single transformation was used to map triangle $A$ onto triangle $B$?

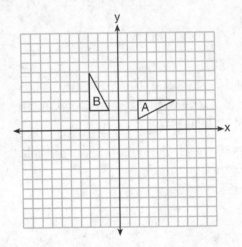

(1) line reflection      (3) dilation

(2) rotation      (4) translation      13 _____

**14** In the diagram below, $\triangle DEF$ is the image of $\triangle ABC$ after a clockwise rotation of $180°$ and a dilation where $AB = 3$, $BC = 5.5$, $AC = 4.5$, $DE = 6$, $FD = 9$, and $EF = 11$.

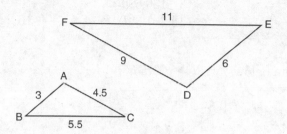

Which relationship must always be true?

(1) $\dfrac{m\angle A}{m\angle D} = \dfrac{1}{2}$      (3) $\dfrac{m\angle A}{m\angle C} = \dfrac{m\angle F}{m\angle D}$

(2) $\dfrac{m\angle C}{m\angle F} = \dfrac{2}{1}$      (4) $\dfrac{m\angle B}{m\angle E} = \dfrac{m\angle C}{m\angle F}$    14 _____

**15** In the diagram below, quadrilateral $ABCD$ is inscribed in circle $P$.

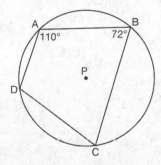

What is $m\angle ADC$?

(1) $70°$      (3) $108°$

(2) $72°$      (4) $110°$      15 _____

**16** A hemispherical tank is filled with water and has a diameter of 10 feet. If water weighs 62.4 pounds per cubic foot, what is the total weight of the water in a full tank, to the *nearest pound*?

(1) 16,336          (3) 130,690
(2) 32,673          (4) 261,381                16 _____

**17** In the diagram below, $\triangle ABC \sim \triangle ADE$.

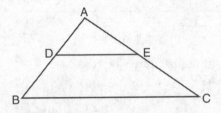

Which measurements are justified by this similarity?

(1) $AD = 3$, $AB = 6$, $AE = 4$, and $AC = 12$
(2) $AD = 5$, $AB = 8$, $AE = 7$, and $AC = 10$
(3) $AD = 3$, $AB = 9$, $AE = 5$, and $AC = 10$
(4) $AD = 2$, $AB = 6$, $AE = 5$, and $AC = 15$          17 _____

**18** Triangle *FGH* is inscribed in circle *O*, the length of radius $\overline{OH}$ is 6, and $\overline{FH} \cong \overline{OG}$.

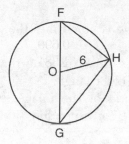

What is the area of the sector formed by angle *FOH*?

(1) $2\pi$          (3) $6\pi$

(2) $\dfrac{3}{2}\pi$          (4) $24\pi$          18 ____

**19** As shown in the diagram below, $\overline{AB}$ and $\overline{CD}$ intersect at *E*, and $\overline{AC} \parallel \overline{BD}$.

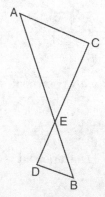

Given $\triangle AEC \sim \triangle BED$, which equation is true?

(1) $\dfrac{CE}{DE} = \dfrac{EB}{EA}$          (3) $\dfrac{EC}{AE} = \dfrac{BE}{ED}$

(2) $\dfrac{AE}{BE} = \dfrac{AC}{BD}$          (4) $\dfrac{ED}{EC} = \dfrac{AC}{BD}$          19 ____

**20** A triangle is dilated by a scale factor of 3 with the center of dilation at the origin. Which statement is true?

(1) The area of the image is nine times the area of the original triangle.

(2) The perimeter of the image is nine times the perimeter of the original triangle.

(3) The slope of any side of the image is three times the slope of the corresponding side of the original triangle.

(4) The measure of each angle in the image is three times the measure of the corresponding angle of the original triangle. 20 _____

**21** The Great Pyramid of Giza was constructed as a regular pyramid with a square base. It was built with an approximate volume of 2,592,276 cubic meters and a height of 146.5 meters. What was the length of one side of its base, to the *nearest meter*?

(1) 73          (3) 133

(2) 77          (4) 230          21 _____

**22** A quadrilateral has vertices with coordinates $(-3, 1)$, $(0, 3)$, $(5, 2)$, and $(-1, -2)$. Which type of quadrilateral is this?

(1) rhombus          (3) square

(2) rectangle          (4) trapezoid          22 _____

**23** In the diagram below, $\triangle ABE$ is the image of $\triangle ACD$ after a dilation centered at the origin. The coordinates of the vertices are $A(0, 0)$, $B(3, 0)$, $C(4.5, 0)$, $D(0, 6)$, and $E(0, 4)$.

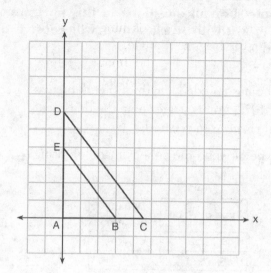

The ratio of the lengths of $\overline{BE}$ to $\overline{CD}$ is

(1) $\dfrac{2}{3}$

(3) $\dfrac{3}{4}$

(2) $\dfrac{3}{2}$

(4) $\dfrac{4}{3}$

23 _____

**24** Line $y = 3x - 1$ is transformed by a dilation with a scale factor of 2 and centered at $(3, 8)$. The line's image is

(1) $y = 3x - 8$

(3) $y = 3x - 2$

(2) $y = 3x - 4$

(4) $y = 3x - 1$

24 _____

## PART II

**Answer all 7 questions in this part. Each correct answer will receive 2 credits. Clearly indicate the necessary steps, including appropriate formula substitutions, diagrams, graphs, charts, etc. For all questions in this part, a correct numerical answer with no work shown will receive only 1 credit.** [14 credits]

**25** A wooden cube has an edge length of 6 centimeters and a mass of 137.8 grams. Determine the density of the cube, to the *nearest thousandth.*

State which type of wood the cube is made of, using the density table below.

| Type of Wood | Density (g/cm³) |
|---|---|
| Pine | 0.373 |
| Hemlock | 0.431 |
| Elm | 0.554 |
| Birch | 0.601 |
| Ash | 0.638 |
| Maple | 0.676 |
| Oak | 0.711 |

**26** Construct an equilateral triangle inscribed in circle $T$ shown below.

[Leave all construction marks.]

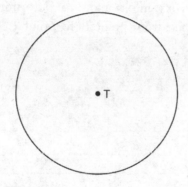

**27** To find the distance across a pond from point $B$ to point $C$, a surveyor drew the diagram below. The measurements he made are indicated on his diagram.

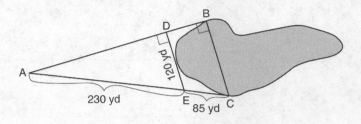

Use the surveyor's information to determine and state the distance from point $B$ to point $C$, to the *nearest yard*.

**28** In parallelogram $ABCD$ shown below, diagonals $\overline{AC}$ and $\overline{BD}$ intersect at $E$.

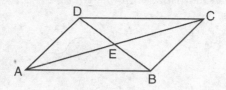

Prove: $\angle ACD \cong \angle CAB$

**29** Triangles $RST$ and $XYZ$ are drawn below. If $RS = 6$, $ST = 14$, $XY = 9$, $YZ = 21$, and $\angle S \cong \angle Y$, is $\triangle RST$ similar to $\triangle XYZ$? Justify your answer.

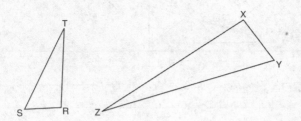

**30** In the diagram below, △*ABC* and △*XYZ* are graphed.

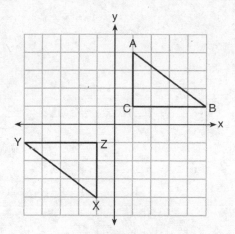

Use the properties of rigid motions to explain why △*ABC* ≅ △*XYZ*.

**31** The endpoints of $\overline{DEF}$ are $D(1, 4)$ and $F(16, 14)$. Determine and state the coordinates of point $E$, if $DE : EF = 2 : 3$.

## PART III

Answer all 3 questions in this part. Each correct answer will receive 4 credits. Clearly indicate the necessary steps, including appropriate formula substitutions, diagrams, graphs, charts, etc. For all questions in this part, a correct numerical answer with no work shown will receive only 1 credit.   [12 credits]

**32** As shown in the diagram below, a ship is heading directly toward a lighthouse whose beacon is 125 feet above sea level. At the first sighting, point *A*, the angle of elevation from the ship to the light was 7°. A short time later, at point *D*, the angle of elevation was 16°.

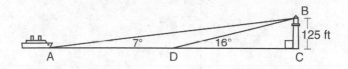

To the *nearest foot*, determine and state how far the ship traveled from point *A* to point *D*.

**33** Triangle *ABC* has vertices with *A*(*x*, 3), *B*(–3, –1), and *C*(–1, –4).

Determine and state a value of *x* that would make triangle *ABC* a right triangle. Justify why △*ABC* is a right triangle.

[The use of the set of axes below is optional.]

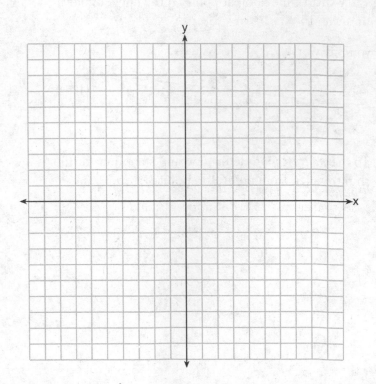

**34** In the diagram below, $\overline{AC} \cong \overline{DF}$ and points $A$, $C$, $D$, and $F$ are collinear on line $\ell$.

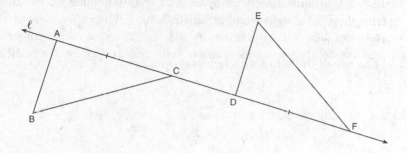

Let $\triangle D'E'F'$ be the image of $\triangle DEF$ after a translation along $\ell$, such that point $D$ is mapped onto point $A$. Determine and state the location of $F'$. Explain your answer.

Let $\triangle D''E''F''$ be the image of $\triangle D'E'F'$ after a reflection across line $\ell$. Suppose that $E''$ is located at $B$. Is $\triangle DEF$ congruent to $\triangle ABC$? Explain your answer.

## PART IV

Answer the 2 questions in this part. Each correct answer will receive 6 credits. Clearly indicate the necessary steps, including appropriate formula substitutions, diagrams, graphs, charts, etc. For all questions in this part, a correct numerical answer with no work shown will receive only 1 credit. [12 credits]

**35** In the diagram of parallelogram *ABCD* below, $\overline{BE} \perp \overline{CED}$, $\overline{DF} \perp \overline{BFC}$, and $\overline{CE} \cong \overline{CF}$.

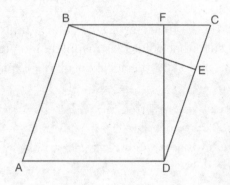

Prove *ABCD* is a rhombus.

**36** Walter wants to make 100 candles in the shape of a cone for his new candle business. The mold shown below will be used to make the candles. Each mold will have a height of 8 inches and a diameter of 3 inches. To the *nearest cubic inch*, what will be the total volume of 100 candles?

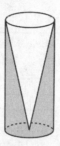

Walter goes to a hobby store to buy the wax for his candles. The wax costs $0.10 per ounce. If the weight of the wax is 0.52 ounce per cubic inch, how much will it cost Walter to buy the wax for 100 candles?

If Walter spent a total of $37.83 for the molds and charges $1.95 for each candle, what is Walter's profit after selling 100 candles?

# Answers
# August 2015
## Geometry (Common Core)

## Answer Key

### PART I

| | | | | | |
|---|---|---|---|---|---|
| **1.** (2) | **5.** (1) | **9.** (3) | **13.** (2) | **17.** (4) | **21.** (4) |
| **2.** (3) | **6.** (4) | **10.** (1) | **14.** (4) | **18.** (3) | **22.** (4) |
| **3.** (4) | **7.** (1) | **11.** (2) | **15.** (3) | **19.** (2) | **23.** (1) |
| **4.** (1) | **8.** (3) | **12.** (3) | **16.** (1) | **20.** (1) | **24.** (4) |

### PART II

**25.** The density of the cube is $0.638$ g/cm$^3$.

**26.** See the construction in the *Answers Explained* section.

**27.** 164 yards

**28.** $\angle ACD$ and $\angle CAB$ are congruent alternate interior angles formed by parallel sides $\overline{AB}$ and $\overline{CD}$.

**29.** The triangles are similar by SAS. The lengths of two pairs of sides are proportional and the included angles are congruent.

**30.** $\triangle XYZ$ is a $180°$ rotation of $\triangle ABC$, and rotations are rigid motions.

**31.** The coordinates of point $E$ are $(7, 8)$.

### PART III

**32.** The ship traveled 582 ft from $A$ to $D$.

**33.** The two possible values of $x$ are 3 and 9.5.

**34.** $F'$ maps to point $C$. $\triangle DEF \cong \triangle ABC$ because a translation followed by a reflection will map one to the other.

### PART IV

**35.** First prove $\triangle CFD \cong \triangle CEB$. Sides $\overline{CD}$ and $\overline{CB}$ are then congruent by CPCTC. $ABCD$ is a rhombus. It is a parallelogram with consecutive congruent sides.

**36.** The volume of 100 candles is 1885 in$^3$, the wax costs \$98.02, and the total profit is \$59.15.

In **PARTS II–IV** you are required to show how you arrived at your answers. For sample methods of solutions, see the *Answers Explained* section.

# Answers Explained

## PART I

1. Sketching a general parallelogram, rhombus, and rectangle is a good strategy for evaluating each of the choices. We need to identify which figure has a property found in the rectangles but not necessarily in parallelograms and rhombuses.

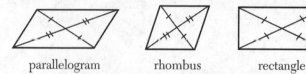

parallelogram          rhombus          rectangle

The rectangle on the right clearly does not have perpendicular diagonals, so choice (1) is not correct. All three figures have congruent opposite sides and parallel opposite sides, so choices (3) and (4) are not correct. The general parallelogram and rhombus each have a long and short diagonal, while the rectangle has two congruent diagonals.

The correct choice is (2).

2. Reflections, translations, and rotations are all rigid motions that result in an image congruent to the pre-image. A dilation with a scale factor of 2 will result in an image larger than the pre-image.

The correct choice is (3).

3. The solid of revolution formed by revolving a rectangle about one of its sides is always a cylinder. The two sides perpendicular to the axis of revolution trace out the two circular bases.

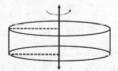

The correct choice is (4).

4. The cofunction relationship states that the sine of an angle is equal to the cosine of its complement. Given angle $x$, the complement of $x$ is $(90° - x)$. Therefore, $\sin x = \cos(90° - x)$.

The correct choice is (1).

**5.** A reflection of a figure over one of its lines of symmetry will always map a figure to itself. Each of the given lines are graphed over the square in the accompanying figure. The only line that is not a line of symmetry is the line $x = 5$.

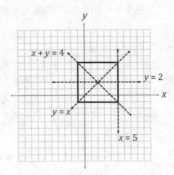

The correct choice is (**1**).

**6.** The dilation will result in a triangle similar to original $\triangle ABC$. Each side is increased in length by a scale factor of $k$, angle measures are preserved, and the image of each side is either parallel or collinear with the pre-image. The important thing to remember as you evaluate each choice in this problem is to not assume information that is not specifically given.

- Choice (1): Even though the scale factor looks to be approximately 2, we do not know that $AD$ is exactly twice $AB$.

- Choice (2): The two triangles look to be right triangles, but we do not know for sure that $\overline{AD} \perp \overline{DE}$.

- Choice (3): $AC$ would be equal to $CE$ only if the scale factor is exactly 2, but we do not know if it is.

- Choice (4): $\overline{DE}$ is the image of $\overline{BC}$ after the dilation, so the two segments must be parallel.

The correct choice is (**4**).

**7.** The first transformation changes the orientation from clockwise in $ABCD$ to counterclockwise in $A'B'C'D'$. Only a reflection changes orientation. We can eliminate choices (3) and (4). The alignment of the figure changes after the second transformation. $A'B'C'D'$ has its long axis aligned, horizontally, while $A''B''C''D''$ has its long axis aligned vertically. A translation is simply a sliding motion of the figure and does not change the alignment, so choice (2) can be eliminated.

The correct choice is (**1**).

**8.** Since opposite angles of a parallelogram are congruent:

$$m\angle EDF = m\angle R$$
$$= 124°$$

Since $\angle EDF$ and $\angle ADF$ are a supplementary linear pair:

$$m\angle ADF + m\angle EDF = 180°$$
$$m\angle ADF + 124° = 180°$$
$$m\angle ADF = 56°$$

Since $\angle ADF$ is isosceles with two congruent base angles:

$$m\angle A = m\angle ADF$$
$$= 56°$$

Finally, from the angle sum theorem we have:

$$m\angle AFD + m\angle A + m\angle ADF = 180°$$
$$m\angle AFD + 56° + 56° = 180°$$
$$m\angle AFD + 112° = 180°$$
$$m\angle AFD = 68°$$

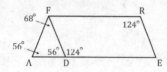

The correct choice is **(3)**.

**9.** Use the completing the square procedure to rewrite the equation in the form $(x - h)^2 + (y - k)^2 = r^2$. First find the constants needed for the $x$-terms and $y$-terms:

$$\text{constant for the } x\text{-terms} = \left(\frac{1}{2} \cdot \text{coefficient of } x\right)^2$$
$$= \left(\frac{1}{2} \cdot 4\right)^2$$
$$= (2)^2$$
$$= 4$$

$$\text{constant for the } y\text{-terms} = \left(\frac{1}{2} \cdot \text{coefficient of } y\right)^2$$
$$= \left(\frac{1}{2} \cdot (-6)\right)^2$$
$$= (-3)^2$$
$$= 9$$

Add 12 to both sides to eliminate the 12 from the left side. Then add the required constants to each side of the original equation and factor:

$$x^2 + 4x + y^2 - 6y - 12 = 0$$
$$x^2 + 4x + y^2 - 6y = 12$$
$$x^2 + 4x + 4 + y^2 - 6y + 9 = 12 + 4 + 9$$
$$(x + 2)^2 + (y - 3)^2 = 25$$

From this equation, $r^2 = 25$ and $r = 5$. The radius of the circle is 5.

The correct choice is **(3)**.

10. Parallel lines have equal slopes, so first calculate the slope of $\overline{MN}$. Two points on the graph are given, so we can find the slope. Use $M(-6, 1)$ as $(x_2, y_2)$ and $N(3, -5)$ as $(x_1, y_1)$:

$$\text{slope} = \frac{y_2 - y_1}{x_2 - x_1}$$
$$= \frac{1 - (-5)}{-6 - 3}$$
$$= \frac{6}{-9}$$
$$= -\frac{2}{3}$$

The desired line will have a slope of $-\frac{2}{3}$. Next find the $y$-intercept by substituting the slope, $-\frac{2}{3}$, and the coordinates for point $P(6, 1)$ into the slope-intercept form of the line:

$$y = mx + b$$
$$1 = \left(-\frac{2}{3}\right)(6) + b$$
$$1 = -4 + b$$
$$b = 5$$

Substitute the slope and $y$-intercept into the slope-intercept form of a line:

$$y = mx + b$$
$$y = -\frac{2}{3}x + 5$$

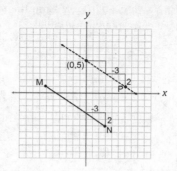

The correct choice is **(1)**.

11. The largest square that can be inscribed in a circle is the square whose diagonals are two perpendicular diameters. The accompanying figure shows perpendicular diameters $\overline{AB}$ and $\overline{CD}$ and square $ABCD$.

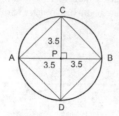

The length of side $\overline{AC}$ is found using the Pythagorean theorem:

$$a^2 + b^2 = c^2$$
$$AP^2 + PC^2 = AC^2$$
$$3.5^2 + 3.5^2 = AC^2$$
$$12.25 + 12.25 = AC^2$$
$$24.5 = AC^2$$
$$AC = \sqrt{24.5}$$
$$AC = 4.949$$
$$AC = 4.9 \text{ inches}$$

The correct choice is **(2)**.

12. This problem involves applying the proportional side relationship in $\triangle BAO$ and $\triangle BCP$, which are similar. $\angle A$ and $\angle C$ are congruent right angles because a radius is always perpendicular to a tangent at the point of tangency. $\angle OBA$ and $\angle CBP$ are congruent vertical angles. Therefore,

$\triangle BAO$ and $\triangle BCP$ are similar by the $AA$ theorem. Calculate the length of $BC$ using proportional corresponding sides:

$$\frac{OA}{PC} = \frac{AB}{BC}$$

$$\frac{4}{10} = \frac{5}{BC}$$

$$4BC = 5 \cdot 10$$

$$4BC = 50$$

$$BC = 12.5$$

The correct choice is (**3**).

13. Start by eliminating choices that are inconsistent with the image.

- Choice (1): Line reflection—In a line reflection, the line of reflection is the perpendicular bisector of the segments joining each pair of corresponding points. Lines $m$ and $n$ join corresponding pairs of points. Line $p$ is perpendicular to $m$ but not $n$. Line reflection is not correct.

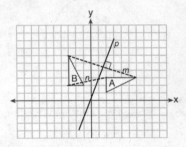

- Choice (3): Dilation—The image and pre-image are congruent, which implies a dilation with a scale factor of 1. However, a dilation with scale factor of 1 would map the pre-image onto itself. Therefore, a dilation is not correct.

- Choice (4): Translation—A translation is a sliding motion that preserves slope, so this cannot be correct.

The transformation must be a rotation.

The correct choice is (**2**).

**14.** $\triangle ABC \sim \triangle DEF$ because all three pairs of corresponding sides are in the same ratio:

$$\frac{AB}{DE} = \frac{3}{6} = \frac{1}{2}$$

$$\frac{AC}{DF} = \frac{4.5}{9} = \frac{1}{2}$$

$$\frac{BC}{EF} = \frac{5.5}{11} = \frac{1}{2}$$

In similar triangles, corresponding angles are always congruent. $m\angle B \cong m\angle E$ because they are congruent corresponding angles, and the ratio $\frac{m\angle B}{m\angle E} = 1$. $m\angle C \cong m\angle F$ for the same reason, so their ratio is also 1. Therefore choice (4) must be true.

The correct choice is **(4)**.

**15.** Quadrilateral $ABCD$ is an inscribed quadrilateral. The opposite angles of inscribed quadrilaterals are always supplementary.

$$m\angle D + m\angle B = 180°$$
$$m\angle D + 72° = 180°$$
$$m\angle D = 108°$$

The correct choice is **(3)**.

**16.** Use the formula for the volume of a sphere (found on the reference sheet). Divide by 2 to calculate the volume of the hemisphere. Then use the relationship

$$\text{weight} = \text{volume} \cdot \text{density}$$

The radius is equal to $\frac{1}{2}$ the diameter:

$$r = \frac{1}{2}d$$

$$r = \frac{1}{2}(10 \text{ ft})$$

$$r = 5 \text{ ft}$$

$$V_{\text{sphere}} = \frac{4}{3}\pi r^3$$

$$V_{\text{hemisphere}} = \frac{1}{2}\left(\frac{4}{3}\pi r^3\right)$$

$$= \frac{1}{2}\left(\frac{4}{3}\pi \cdot 5^3\right)$$

$$= 261.7993 \text{ ft}^3$$

$$\text{weight} = \text{density} \cdot \text{volume}$$

$$= 261.7993 \text{ ft}^3 \cdot 62.4 \, \frac{\text{lb}}{\text{ft}^3}$$

$$= 16{,}336.2817 \text{ lb}$$

$$= 16{,}336 \text{ lb}$$

The correct choice is **(1)**.

17. Corresponding parts of similar triangles are proportional. The sides listed in the choices must satisfy the proportion $\dfrac{AD}{AB} = \dfrac{AE}{AC}$. Evaluate each choice:

- Choice (1): $\quad\dfrac{3}{6} = \dfrac{4}{12}$

  $\qquad 3 \cdot 12 = 6 \cdot 4 \qquad$ Cross-multiply to check

  $\qquad 36 \neq 24$

- Choice (2): $\quad\dfrac{5}{8} = \dfrac{7}{10}$

  $\qquad 8 \cdot 7 = 5 \cdot 10 \qquad$ Cross-multiply to check

  $\qquad 56 \neq 50$

- Choice (3): $\quad\dfrac{3}{9} = \dfrac{5}{10}$

  $\qquad 3 \cdot 10 = 9 \cdot 5 \qquad$ Cross-multiply to check

  $\qquad 40 \neq 45$

- Choice (4): $\quad\dfrac{2}{6} = \dfrac{5}{15}$

  $\qquad 2 \cdot 15 = 6 \cdot 5 \qquad$ Cross-multiply to check

  $\qquad 30 = 30$

The correct choice is **(4)**.

**18.** $\overline{OF} \cong \overline{OH} \cong \overline{OG}$ because radii of the same circle are all congruent. $\overline{FH} \cong \overline{OG}$ is given. By substitution, $\overline{OF} \cong \overline{OH} \cong \overline{FH}$. Therefore $\triangle FOH$ is equilateral because all of its sides are congruent. All angles in an equilateral triangle measure 60°, so m$\angle FOH$ = 60°. The area of the sector can now be calculated. The central angle measures 60°, and the radius is 6. The area of a sector equals the area of the circle multiplied by the central angle divided by 360°.

$$A_{\text{sector } FOH} = \frac{\text{m}\angle FOH}{360°} \pi r^2$$

$$= \frac{60°}{360°} \pi (6)^2$$

$$= 6\pi$$

The correct choice is **(3)**.

**19.** The similarity statement $\triangle AEC \sim \triangle BED$, tells us that corresponding pairs of sides are proportional. Examine each choice to identify the one that properly matches up corresponding sides.

- Choice (1): $\dfrac{CE}{DE} = \dfrac{EB}{EA}$ involves two ratios formed from corresponding parts. However, the ratio on the left has the side from the larger triangle in the numerator while the ratio on the right has the side from the larger triangle in the denominator. This ratio is not correct.

- Choice (2): $\dfrac{AE}{BE} = \dfrac{AC}{BD}$ has two ratios formed from corresponding parts, and the side from the larger triangle appears in the two numerators. This choice is correct.

- Choice (3): $\dfrac{EC}{AE} = \dfrac{BE}{ED}$ has corresponding sides incorrectly matched. $\overline{EC}$ corresponds to $\overline{ED}$ and $\overline{AE}$ corresponds to $\overline{BE}$. This choice is not correct.

- Choice (4): $\dfrac{ED}{EC} = \dfrac{AC}{BD}$ is also incorrect for the same reasons as in choice (3). $\overline{ED}$ corresponds to $\overline{BD}$, not $\overline{AC}$.

The correct choice is **(2)**.

**20.** After a dilation, the ratio of the areas is equal to the scale factor squared, or 9 in this case. Therefore, the area of the image equals 9 times the area of the pre-image. Choice (1) is correct. The ratio of the perimeters is equal to the scale factor, or 3, so choice (2) is not correct. A dilation always preserves slope and angle, so choices (3) and (4) are not correct.

The correct choice is **(1)**.

**21.** Use the formula for the volume of a pyramid from the reference sheet to find the area of the square base. The side length of the square can then be calculated from its area.

$V = \dfrac{1}{3} Bh$, where $B$ is the area of the square base and $h$ is the height of the pyramid.

$$2{,}592{,}276 \text{ m}^3 = \dfrac{1}{3} B(146.5 \text{ m})$$

$$7{,}776{,}828 \text{ m}^3 = B(146.5 \text{ m})$$

$$B = 53{,}084.1501 \text{ m}^2$$

The square base has an area of $53{,}084.1501 \text{ m}^2$.

$$A_{\text{square}} = s^2$$

$$53{,}084.1501 \text{ m}^2 = s^2$$

$$s = 230.39 \text{ m}$$

$$s = 230 \text{ m}$$

The correct choice is **(4)**.

**22.** It is a good idea to start by graphing the quadrilateral, as shown in the accompanying figure. The figure has a pair of opposite sides that are clearly not parallel. The first three choices can be eliminated because the rhombus, rectangle, and square are types of parallelograms, which always have *two* pairs of parallel sides. The figure is a trapezoid, which has *at least one* pair of parallel sides.

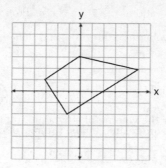

The correct choice is **(4)**.

**23.** One approach would be to use the distance formula or Pythagorean theorem to calculate the lengths $BE$ and $CD$. However, a shortcut is to find the ratio of $EA$ to $DA$. Dilations result in similar figures, and all pairs of corresponding sides have the same ratio. Count the number of vertical boxes on the grid to find the lengths:

$$EA = 4$$
$$DA = 6$$
$$\frac{EA}{DA} = \frac{4}{6}$$
$$\frac{EA}{DA} = \frac{2}{3}$$

The correct choice is **(1)**.

**24.** Check to see if the center of dilation lies on the given line. If so, the line will map to itself. If the point $(3, 8)$ satisfies the equation of the line, then it lies on the line:

$$y = 3x - 1$$
$$8 = 3(3) - 1$$
$$8 = 9 - 1$$
$$8 = 8$$

The point $(3, 8)$ lies on the given line, so the line will map to itself.

The correct choice is **(4)**.

# PART II

**25.** First calculate the volume of the cube, and then use the volume to calculate the density.

$$V_{cube} = s^3$$
$$= (6 \text{ cm})^3$$
$$= 216 \text{ cm}^3$$

Volume and density are related by the formula density $= \dfrac{\text{mass}}{\text{volume}}$:

$$\text{density} = \dfrac{\text{mass}}{\text{volume}}$$

$$\text{density} = \dfrac{137.8 \text{ g}}{216 \text{ cm}^3}$$

$$\text{density} = 0.6379 \text{ g/cm}^3$$

$$\text{density} = 0.638 \text{ g/cm}^3$$

From the table, we see that ash has a density of $0.638 \text{ g/cm}^3$.

The density of the cube, to the *nearest thousandth*, is **0.638 g/cm³**. The cube is made of ash.

**26.** Construct an inscribed equilateral triangle as follows:

1. Mark point $A$ at any location on the circle.

2. Place the compass point at $T$ and the pencil at $A$ to measure the radius of the circle.

3. Without changing the compass opening, place the compass point at $A$ and make an arc that intersects the circle at $B$.

4. With the same compass opening, place the compass point at $B$ and make arc that intersects the circle at $C$. Continue making arcs intersecting at $D$, $E$, and $F$.

5. Use a straightedge to connect point $A$ to point $C$, point $C$ to point $E$, and point $E$ to point $A$.

6. $\triangle ACE$ is an equilateral triangle.

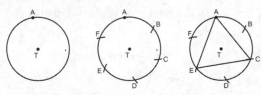

**27.** From the figure, we can show that $\triangle ADE$ and $\triangle ABC$ are similar. Shared angle $\angle A$ is congruent to itself, and $\angle B \cong \angle D$ because they are both right angles. Therefore, $\triangle ADE \sim \triangle ABC$ by the AA theorem. We can use the fact that corresponding sides are proportional to write a proportion that will let us solve for the length $BC$:

$$\frac{AE}{AC} = \frac{DE}{BC}$$

$$\frac{230}{230 + 85} = \frac{120}{BC} \qquad AC \text{ is equal to } AE + EC$$

$$\frac{230}{315} = \frac{120}{BC}$$

$$230BC = 120 \cdot 315 \qquad \text{Cross-multiply}$$

$$230BC = 37{,}800$$

$$BC = 164.34$$

$$BC = 164 \text{ yd}$$

To the *nearest yard*, the distance from point $B$ to point $C$ is **164** yards.

**28.** It is given that $ABCD$ is a parallelogram. $\overline{AB}$ // $\overline{CD}$ because opposite sides of a parallelogram are parallel. From the figure, transversal $\overline{AC}$ forms alternate interior angles $\angle ACD$ and $\angle CAB$. $\angle ACD \cong \angle CAB$ because a transversal intersecting parallel lines forms congruent alternate interior angles.

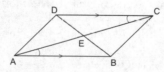

**29.** Start by marking the congruent pairs of sides and angles in the figure.

Dimensions are given for two pairs of sides, and the included angles are congruent. This suggests the SAS similarity theorem. Check that the lengths of the two pairs of sides are proportional:

$$\frac{RS}{ST} = \frac{XY}{YZ}$$

$$\frac{6}{14} = \frac{9}{21}$$

$$9 \cdot 14 = 6 \cdot 21 \qquad \text{Cross-multiply to confirm the proportion}$$

$$126 = 126$$

The triangles are similar by the SAS similarity theorem because the lengths of two pairs of sides are proportional and the included angles are congruent.

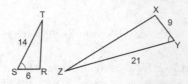

**30.** $\triangle XYZ$ is the image of $\triangle ABC$ after a 180° rotation about the origin. Rotations are rigid motions that preserve length and angle measure. So the pre-image and image are congruent. Therefore, $\triangle ABC \cong \triangle XYZ$.

**31.** Find the $x$- and $y$-coordinates of point $E$ using the formulas:

$$\text{ratio} = \frac{x - x_1}{x_2 - x}$$

$$\text{ratio} = \frac{y - y_1}{y_2 - y}$$

Using the coordinates $D(1, 4)$ and $F(16, 14)$, $x_1 = 1$, $x_2 = 16$, $y_1 = 4$, and $y_2 = 14$. The ratio is $\frac{2}{3}$.

Solve for the $x$-coordinate:

$$\text{ratio} = \frac{x - x_1}{x_2 - x}$$

$$\frac{2}{3} = \frac{x-1}{16-x}$$

$$3(x-1) = 2(16-x) \qquad \text{Cross-multiply}$$

$$3x - 3 = 32 - 2x$$

$$5x - 3 = 32$$

$$5x = 35$$

$$x = 7$$

Solve for the $y$-coordinate:

$$\text{ratio} = \frac{y - x_1}{x_2 - y}$$

$$\frac{2}{3} = \frac{y-4}{14-y}$$

$$3(y-4) = 2(14-y)$$

$$3y - 12 = 28 - 2y$$

$$5y - 12 = 28$$

$$5y = 40$$

$$y = 8$$

The coordinates of point $E$ are **(7, 8)**.

## PART III

**32.** From the diagram, $AD$ is equal to $AC - DC$. The lengths of $AC$ and $DC$ can be found by applying trigonometry to $\triangle ACB$ and $\triangle DCB$, respectively.

First find $AC$ using $\triangle ACB$. Relative to $\angle A$, $AC$ is the adjacent side and $BC$ is the opposite side. Use the tangent ratio:

$$\tan = \frac{\text{opposite}}{\text{adjacent}}$$

$$\tan(A) = \frac{BC}{AC}$$

$$\tan(7°) = \frac{125}{AC}$$

$$AC \tan(7°) = 125$$

$$AC = \frac{125}{\tan(7°)}$$

$$AC = 1{,}018.04 \text{ ft}$$

Next find $DC$ using $\triangle DCB$. Relative to $\angle A$, $DC$ is the adjacent side and $BC$ is the opposite side. Use the tangent ratio:

$$\tan = \frac{\text{opposite}}{\text{adjacent}}$$

$$\tan(D) = \frac{BC}{DC}$$

$$\tan(16°) = \frac{125}{DC}$$

$$DC \tan(16°) = 125$$

$$DC = \frac{125}{\tan(16°)}$$

$$DC = 435.92 \text{ ft}$$

Now calculate $AD$:

$$AD = AC - DC$$
$$AD = 1{,}018.04 \text{ ft} - 435.92 \text{ ft}$$
$$AD = 582.12 \text{ ft}$$
$$AD = 582 \text{ ft}$$

To the *nearest foot*, the ship traveled **582** feet from $A$ to $D$.

**33.** Begin by graphing the points $B(-3, -1)$ and $C(-1, -4)$. Point $A$ can be located so that any of the angles is a right angle. We will choose $\angle B$ to be the right angle.

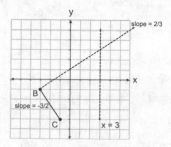

The slopes of perpendicular lines are negative reciprocals. Calculate the slope of $\overline{BC}$, and then find the slope of $\overline{AB}$ by taking the negative reciprocal:

$$\text{slope of } \overline{BC} = \frac{y_2 - y_1}{x_2 - x_1}$$

$$= \frac{-4 - (-1)}{-1 - (-3)}$$

$$= \frac{-3}{2}$$

$$= -\frac{3}{2}$$

The slope of $\overline{AB}$ is the negative reciprocal, or $\frac{2}{3}$. Graph a line through point $B$ with a slope of $\frac{2}{3}$. Point $A$ will lie on this line. We know that the $x$-coordinate of point $A$ is 3, so graph the vertical line $x = 3$ on the same graph. Point $A$ is located at the intersection of these two lines.

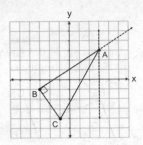

The coordinates of *A* are (3, 3).

The value of *x* is **3**. The procedure is the same if point *C* is chosen as the vertex of the right angle. The resulting value of *x* is 9.5. Either *x* = 3 or *x* = 9.5 is acceptable.

**34.** $\overline{AC} \cong \overline{DF}$, so a translation along line $\ell$ that maps point *D* to *A* must map point *F* to *C*.

Therefore point *F′* will be located at **point C**.

*D″* maps to *A*, and *F″* maps to *C* because points on the line of reflection are unchanged after a reflection. We are given *E″* is located at *B*. Each point on $\triangle D''E''F''$ lies on a vertex of $\triangle ABC$, so $\triangle D''E''F'' \cong \triangle ABC$. We also know $\triangle D''E''F'' \cong \triangle DEF$ because translations and reflections are rigid motions.

Therefore **$\triangle DEF \cong \triangle ABC$**.

## PART IV

**35.** The strategy is first prove $\triangle CFD \cong CEB$. Sides $\overline{CD}$ and $\overline{CB}$ are then congruent by CPCTC. We can then apply the fact that a parallelogram with two consecutive congruent sides is a rhombus.

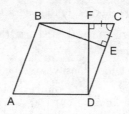

| Statement | Reason |
|---|---|
| 1. $\overline{BE} \perp \overline{CED}$, $\overline{DF} \perp \overline{BFC}$ | 1. Given |
| 2. $\angle CFD$ and $\angle CEB$ are right angles | 2. Perpendicular lines intersect at right angles |
| 3. $\angle CFD \cong \angle CEB$ | 3. All right angles are congruent |
| 4. $\overline{CE} \cong \overline{CF}$ | 4. Given |
| 5. $\angle C \cong \angle C$ | 5. Reflexive property |
| 6. $\triangle CFD \cong \triangle CEB$ | 6. ASA |
| 7. $\overline{CD} \cong \overline{CB}$ | 7. CPCTC |
| 8. $ABCD$ is a parallelogram | 8. Given |
| 9. $ABCD$ is a rhombus | 9. A parallelogram with two consecutive congruent sides is a rhombus |

**36.** Find the volume of each candle using the formula for the volume of a cone from the reference sheet. The height is 8 inches. The radius is $\frac{1}{2}$ the diameter, or 1.5 inches.

$$V = \frac{1}{3}\pi r^2 h$$

$$= \frac{1}{3}\pi(1.5 \text{ in})^2(8 \text{ in})$$

$$= 18.8495 \text{ in}^3$$

Multiply by 100 to find the volume of 100 candles, and round to the nearest cubic inch.

$$V_{100 \text{ candles}} = 100(18.8495 \text{ in}^3)$$
$$= 1,885 \text{ in}^3$$

The total volume of 100 candles, to the *nearest cubic inch*, is **1,885**.

To calculate the cost of the wax, first use the density to calculate the weight of 100 candles:

$$\text{weight} = \text{density} \cdot \text{volume}$$
$$= 0.52 \, \frac{\text{ounce}}{\text{in}^3} \cdot 1,885 \text{ in}^3$$
$$= 980.2 \text{ ounces}$$

Find the cost by multiplying the cost per ounce by the weight:

$$\text{cost} = \frac{\$0.10}{\text{ounce}} \cdot 980.2 \text{ ounces}$$

The cost of the wax for 100 candles is **$98.02**.

Now calculate the total cost to make 100 candles:

$$\text{total cost to make candles} = \text{cost for the molds} + \text{cost for the wax}$$
$$= \$37.83 + \$98.02$$
$$= \$135.85$$

The profit is the difference between the selling price of all the candles and the total cost for all the candles:

$$\text{profit} = \text{total selling price} - \text{total cost to make candles}$$
$$= \frac{\$1.95}{\text{candle}} \cdot 100 \text{ candles} - \$135.85$$
$$= \$195 - \$135.85$$
$$= \$59.15$$

The total profit for selling 100 candles is **$59.15**.

| Topic | Question Numbers | Number of Points | Your Points | Your Percentage |
|---|---|---|---|---|
| 1. Basic Angle and Segment Relationships | | 0 | | |
| 2. Angle and Segment Relationships in Triangles and Polygons | 8, 11 | 2 + 2 = 4 | | |
| 3. Constructions | 26 | 2 | | |
| 4. Transformations | 5, 6, 7, 13, 20, 23, 24, 34 | 2 + 2 + 2 + 2 + 2 + 2 + 2 + 4 = 18 | | |
| 5. Triangle Congruence | 2, 30, 34 | 2 + 2 + 4 = 8 | | |
| 6. Line, Segments and Circles on the Coordinate Plane | 10, 14, 31 | 2 + 2 + 2 = 6 | | |
| 7. Similarity | 12, 14, 17, 19, 27, 29 | 2 + 2 + 2 + 2 + 2 + 2 = 12 | | |
| 8. Trigonometry | 4, 32 | 2 + 4 = 6 | | |
| 9. Parallelograms | 1, 8, 28, 35 | 2 + 2 + 2 + 6 = 12 | | |
| 10. Coordinate Geometry Proofs | 22, 33 | 2 + 4 = 6 | | |
| 11. Volume | 3, 21 | 2 + 2 = 4 | | |
| 12. Modeling | 16, 25, 36 | 2 + 2 + 6 = 10 | | |

# HOW TO CONVERT YOUR RAW SCORE TO YOUR GEOMETRY REGENTS EXAMINATION SCORE

The conversion chart below can be used to determine your final score on the August 2015 Regents Exam in Geometry. To find your final exam score, locate in the "Raw Score" column the total number of points you scored out of a possible 86. Then locate in the adjacent column to the right the scale score that corresponds to your raw score. The scale score is your final score.

| Raw Score | Scale Score | Performance Level | Raw Score | Scale Score | Performance Level | Raw Score | Scale Score | Performance Level |
|---|---|---|---|---|---|---|---|---|
| 86 | 100 | 5 | 57 | 80 | 4 | 28 | 59 | 2 |
| 85 | 99 | 5 | 56 | 80 | 4 | 27 | 58 | 2 |
| 84 | 98 | 5 | 55 | 79 | 3 | 26 | 56 | 2 |
| 83 | 97 | 5 | 54 | 79 | 3 | 25 | 55 | 2 |
| 82 | 96 | 5 | 53 | 79 | 3 | 24 | 53 | 1 |
| 81 | 95 | 5 | 52 | 78 | 3 | 23 | 52 | 1 |
| 80 | 94 | 5 | 51 | 78 | 3 | 22 | 50 | 1 |
| 79 | 93 | 5 | 50 | 77 | 3 | 21 | 49 | 1 |
| 78 | 93 | 5 | 49 | 77 | 3 | 20 | 47 | 1 |
| 77 | 92 | 5 | 48 | 76 | 3 | 19 | 45 | 1 |
| 76 | 91 | 5 | 47 | 75 | 3 | 18 | 44 | 1 |
| 75 | 90 | 5 | 46 | 75 | 3 | 17 | 42 | 1 |
| 74 | 90 | 5 | 45 | 74 | 3 | 16 | 40 | 1 |
| 73 | 89 | 5 | 44 | 74 | 3 | 15 | 38 | 1 |
| 72 | 88 | 5 | 43 | 73 | 3 | 14 | 36 | 1 |
| 71 | 88 | 5 | 42 | 72 | 3 | 13 | 34 | 1 |
| 70 | 87 | 5 | 41 | 72 | 3 | 12 | 32 | 1 |
| 69 | 86 | 5 | 40 | 71 | 3 | 11 | 30 | 1 |
| 68 | 86 | 5 | 39 | 70 | 3 | 10 | 27 | 1 |
| 67 | 86 | 5 | 38 | 69 | 3 | 9 | 25 | 1 |
| 66 | 85 | 5 | 37 | 69 | 3 | 8 | 23 | 1 |
| 65 | 84 | 4 | 36 | 68 | 3 | 7 | 20 | 1 |
| 64 | 84 | 4 | 35 | 67 | 3 | 6 | 18 | 1 |
| 63 | 83 | 4 | 34 | 66 | 3 | 5 | 15 | 1 |
| 62 | 83 | 4 | 33 | 65 | 3 | 4 | 12 | 1 |
| 61 | 82 | 4 | 32 | 64 | 2 | 3 | 10 | 1 |
| 60 | 82 | 4 | 31 | 63 | 2 | 2 | 7 | 1 |
| 59 | 81 | 4 | 30 | 61 | 2 | 1 | 3 | 1 |
| 58 | 81 | 4 | 29 | 60 | 2 | 0 | 0 | 1 |

# Examination June 2016
## Geometry (Common Core)

## GEOMETRY REFERENCE SHEET

| | |
|---|---|
| 1 inch = 2.54 centimeters | 1 ton = 2000 pounds |
| 1 meter = 39.37 inches | 1 cup = 8 fluid ounces |
| 1 mile = 5280 feet | 1 pint = 2 cups |
| 1 mile = 1760 yards | 1 quart = 2 pints |
| 1 mile = 1.609 kilometers | 1 gallon = 4 quarts |
| 1 kilometer = 0.62 mile | 1 gallon = 3.785 liters |
| 1 pound = 16 ounces | 1 liter = 0.264 gallon |
| 1 pound = 0.454 kilogram | 1 liter = 1000 cubic centimeters |
| 1 kilogram = 2.2 pounds | |

| | |
|---|---|
| Triangle | $A = \dfrac{1}{2}bh$ |
| Parallelogram | $A = bh$ |
| Circle | $A = \pi r^2$ |
| Circle | $C = \pi d$ or $C = 2\pi r$ |
| General Prisms | $V = Bh$ |
| Cylinder | $V = \pi r^2 h$ |
| Sphere | $V = \dfrac{4}{3}\pi r^3$ |

| | |
|---|---|
| Cone | $V = \frac{1}{3}\pi r^2 h$ |
| Pyramid | $V = \frac{1}{3}Bh$ |
| Pythagorean Theorem | $a^2 + b^2 = c^2$ |
| Quadratic Formula | $x = \dfrac{-b \pm \sqrt{b^2 - 4ac}}{2a}$ |
| Arithmetic Sequence | $a_n = a_1 + (n-1)d$ |
| Geometric Sequence | $a_n = a_1 r^{n-1}$ |
| Geometric Series | $S_n = \dfrac{a_1 - a_1 r^n}{1 - r}$ where $r \neq 1$ |
| Radians | 1 radian = $\dfrac{180}{\pi}$ degrees |
| Degrees | 1 degree = $\dfrac{\pi}{180}$ radians |
| Exponential Growth/Decay | $A = A_0 e^{k(t - t_0)} + B_0$ |

## PART I

Answer all 24 questions in this part. Each correct answer will receive 2 credits. No partial credit will be allowed. For each statement or question, write in the space provided the numeral preceding the word or expression that best completes the statement or answers the question.  [48 credits]

**1** A student has a rectangular postcard that he folds in half lengthwise. Next, he rotates it continuously about the folded edge. Which three-dimensional object below is generated by this rotation?

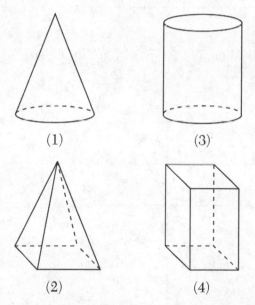

(1)                    (3)

(2)                    (4)              1 _____

**2** A three-inch line segment is dilated by a scale factor of 6 and centered at its midpoint. What is the length of its image?

(1) 9 inches              (3) 15 inches

(2) 2 inches              (4) 18 inches        2 _____

**3** Kevin's work for deriving the equation of a circle is shown below.

$$x^2 + 4x = -(y^2 - 20)$$

STEP 1   $x^2 + 4x = -y^2 + 20$

STEP 2   $x^2 + 4x + 4 = -y^2 + 20 - 4$

STEP 3   $(x + 2)^2 = -y^2 + 20 - 4$

STEP 4   $(x + 2)^2 + y^2 = 16$

In which step did he make an error in his work?

(1) Step 1          (3) Step 3
(2) Step 2          (4) Step 4          3 _____

**4** Which transformation of $\overline{OA}$ would result in an image parallel to $\overline{OA}$?

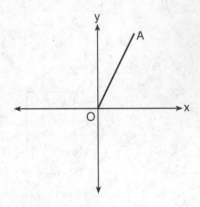

(1) a translation of two units down
(2) a reflection over the x-axis
(3) a reflection over the y-axis
(4) a clockwise rotation of 90° about the origin          4 _____

**5** Using the information given below, which set of triangles can *not* be proven similar?

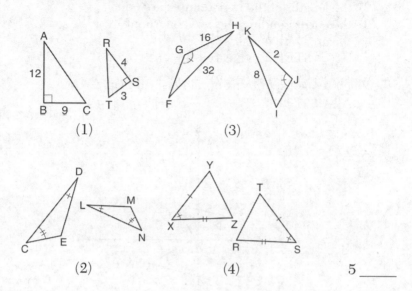

(1)　　　　　　　　　　　　　(3)

(2)　　　　　　　　　　　　　(4)　　　　　　5 _____

**6** A company is creating an object from a wooden cube with an edge length of 8.5 cm. A right circular cone with a diameter of 8 cm and an altitude of 8 cm will be cut out of the cube. Which expression represents the volume of the remaining wood?

(1) $(8.5)^3 - \pi(8)^2(8)$　　　　(3) $(8.5)^3 - \frac{1}{3}\pi(8)^2(8)$

(2) $(8.5)^3 - \pi(4)^2(8)$　　　　(4) $(8.5)^3 - \frac{1}{3}\pi(4)^2(8)$　　　　6 _____

**7** Two right triangles must be congruent if

   (1) an acute angle in each triangle is congruent
   (2) the lengths of the hypotenuses are equal
   (3) the corresponding legs are congruent
   (4) the areas are equal              7 _____

**8** Which sequence of transformations will map $\triangle ABC$ onto $\triangle A'B'C'$?

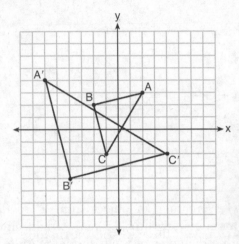

   (1) reflection and translation
   (2) rotation and reflection
   (3) translation and dilation
   (4) dilation and rotation            8 _____

**9** In parallelogram $ABCD$, diagonals $\overline{AC}$ and $\overline{BD}$ intersect at $E$. Which statement does *not* prove parallelogram $ABCD$ is a rhombus?

   (1) $\overline{AC} \cong \overline{DB}$
   (2) $\overline{AB} \cong \overline{BC}$
   (3) $\overline{AC} \perp \overline{DB}$
   (4) $\overline{AC}$ bisects $\angle DCB$.         9 _____

**10** In the diagram below of circle $O$, $\overline{OB}$ and $\overline{OC}$ are radii, and chords $\overline{AB}$, $\overline{BC}$, and $\overline{AC}$ are drawn.

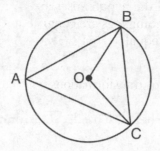

Which statement must always be true?

(1) $\angle BAC \cong \angle BOC$

(2) $m\angle BAC = \dfrac{1}{2} m\angle BOC$

(3) $\triangle BAC$ and $\triangle BOC$ are isosceles.

(4) The area of $\triangle BAC$ is twice the area of $\triangle BOC$.     10 _____

**11** A 20-foot support post leans against a wall, making a 70° angle with the ground. To the *nearest tenth of a foot*, how far up the wall will the support post reach?

(1) 6.8          (3) 18.7

(2) 6.9          (4) 18.8          11 _____

**12** Line segment $NY$ has endpoints $N(-11, 5)$ and $Y(5, -7)$. What is the equation of the perpendicular bisector of $\overline{NY}$?

(1) $y + 1 = \dfrac{4}{3}(x + 3)$      (3) $y - 6 = \dfrac{4}{3}(x - 8)$

(2) $y + 1 = -\dfrac{3}{4}(x + 3)$      (4) $y - 6 = -\dfrac{3}{4}(x - 8)$      12 _____

**13** In $\triangle RST$ shown below, altitude $\overline{SU}$ is drawn to $\overline{RT}$ at $U$.

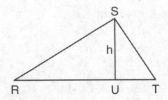

If $SU = h$, $UT = 12$, and $RT = 42$, which value of $h$ will make $\triangle RST$ a right triangle with $\angle RST$ as a right angle?

(1) $6\sqrt{3}$      (3) $6\sqrt{14}$

(2) $6\sqrt{10}$      (4) $6\sqrt{35}$      13 _____

**14** In the diagram below, $\triangle ABC$ has vertices $A(4, 5)$, $B(2, 1)$, and $C(7, 3)$.

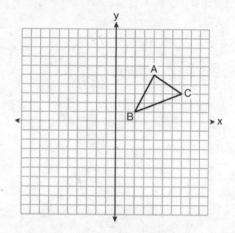

What is the slope of the altitude drawn from $A$ to $\overline{BC}$?

(1) $\dfrac{2}{5}$

(3) $-\dfrac{1}{2}$

(2) $\dfrac{3}{2}$

(4) $-\dfrac{5}{2}$

14 _____

**15** In the diagram below, $\triangle ERM \sim \triangle JTM$.

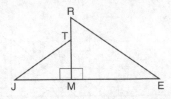

Which statement is always true?

(1) $\cos J = \dfrac{RM}{RE}$          (3) $\tan T = \dfrac{RM}{EM}$

(2) $\cos R = \dfrac{JM}{JT}$          (4) $\tan E = \dfrac{TM}{JM}$          15 _____

**16** On the set of axes below, rectangle $ABCD$ can be proven congruent to rectangle $KLMN$ using which transformation?

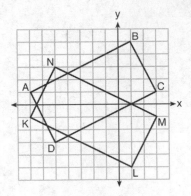

(1) rotation
(2) translation
(3) reflection over the $x$-axis
(4) reflection over the $y$-axis          16 _____

**17** In the diagram below, $\overline{DB}$ and $\overline{AF}$ intersect at point $C$, and $\overline{AD}$ and $\overline{FBE}$ are drawn.

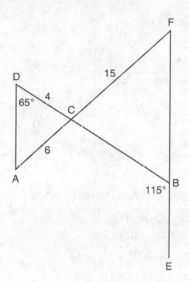

If $AC = 6$, $DC = 4$, $FC = 15$, $m\angle D = 65°$, and $m\angle CBE = 115°$, what is the length of $\overline{CB}$?

(1) 10          (3) 17

(2) 12          (4) 22.5        17 _____

**18** Seawater contains approximately 1.2 ounces of salt per liter on average. How many gallons of seawater, to the *nearest tenth of a gallon*, would contain 1 pound of salt?

(1) 3.3          (3) 4.7

(2) 3.5          (4) 13.3        18 _____

**19** Line segment *EA* is the perpendicular bisector of $\overline{ZT}$, and $\overline{ZE}$ and $\overline{TE}$ are drawn.

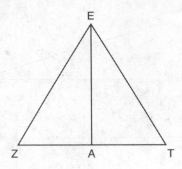

Which conclusion can *not* be proven?

(1) $\overline{EA}$ bisects angle *ZET*.
(2) Triangle *EZT* is equilateral.
(3) $\overline{EA}$ is a median of triangle *EZT*.
(4) Angle *Z* is congruent to angle *T*.          19 _____

**20** A hemispherical water tank has an inside diameter of 10 feet. If water has a density of 62.4 pounds per cubic foot, what is the weight of the water in a full tank, to the *nearest pound*?

(1) 16,336          (3) 130,690
(2) 32,673          (4) 261,381          20 _____

**21** In the diagram of $\triangle ABC$, points $D$ and $E$ are on $\overline{AB}$ and $\overline{CB}$, respectively, such that $\overline{AC} \parallel \overline{DE}$.

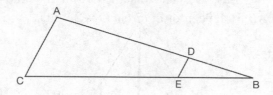

If $AD = 24$, $DB = 12$, and $DE = 4$, what is the length of $\overline{AC}$ ?

(1) 8                  (3) 16

(2) 12                 (4) 72                          21 _____

**22** Triangle $RST$ is graphed on the set of axes below.

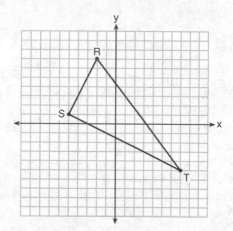

How many square units are in the area of $\triangle RST$?

(1) $9\sqrt{3} + 15$            (3) 45

(2) $9\sqrt{5} + 15$            (4) 90                22 _____

**23** The graph below shows $\overline{AB}$ which is a chord of circle $O$. The coordinates of the endpoints of $\overline{AB}$ are $A(3, 3)$ and $B(3, -7)$. The distance from the midpoint of $\overline{AB}$ to the center of circle $O$ is 2 units.

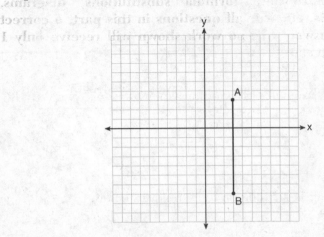

What could be a correct equation for circle $O$?

(1) $(x - 1)^2 + (y + 2)^2 = 29$
(2) $(x + 5)^2 + (y - 2)^2 = 29$
(3) $(x - 1)^2 + (y - 2)^2 = 25$
(4) $(x - 5)^2 + (y + 2)^2 = 25$

23 _____

**24** What is the area of a sector of a circle with a radius of 8 inches and formed by a central angle that measures 60°?

(1) $\dfrac{8\pi}{3}$      (3) $\dfrac{32\pi}{3}$

(2) $\dfrac{16\pi}{3}$      (4) $\dfrac{64\pi}{3}$

24 _____

## PART II

Answer all 7 questions in this part. Each correct answer will receive 2 credits. Clearly indicate the necessary steps, including appropriate formula substitutions, diagrams, graphs, charts, etc. For all questions in this part, a correct numerical answer with no work shown will receive only 1 credit.   [14 credits]

**25** Describe a sequence of transformations that will map △ABC onto △DEF as shown below.

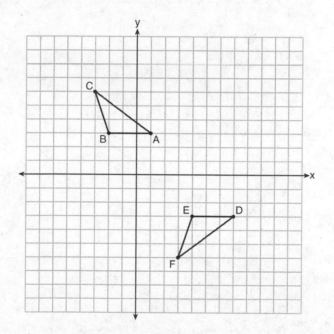

**26** Point $P$ is on segment $AB$ such that $AP : PB$ is $4 : 5$. If $A$ has coordinates $(4, 2)$, and $B$ has coordinates $(22, 2)$, determine and state the coordinates of $P$.

**27** In $\triangle CED$ as shown below, points $A$ and $B$ are located on sides $\overline{CE}$ and $\overline{ED}$, respectively. Line segment $AB$ is drawn such that $AE = 3.75$, $AC = 5$, $EB = 4.5$, and $BD = 6$.

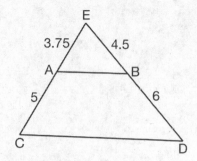

Explain why $\overline{AB}$ is parallel to $\overline{CD}$.

**28** Find the value of $R$ that will make the equation sin 73° = cos $R$ true when $0° < R < 90°$.

Explain your answer.

**29** In the diagram below, Circle 1 has radius 4, while Circle 2 has radius 6.5. Angle $A$ intercepts an arc of length $\pi$, and angle $B$ intercepts an arc of length $\dfrac{13\pi}{8}$.

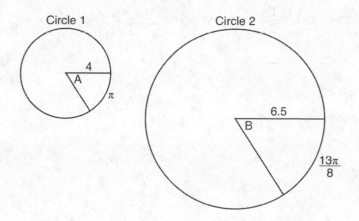

Dominic thinks that angles $A$ and $B$ have the same radian measure. State whether Dominic is correct or not. Explain why.

**30** A ladder leans against a building. The top of the ladder touches the building 10 feet above the ground. The foot of the ladder is 4 feet from the building. Find, to the *nearest degree*, the angle that the ladder makes with the level ground.

**31** In the diagram below, radius $\overline{OA}$ is drawn in circle $O$. Using a compass and a straightedge, construct a line tangent to circle $O$ at point $A$. [Leave all construction marks.]

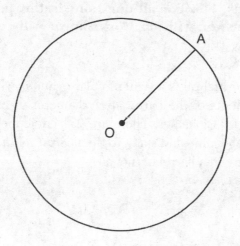

## PART III

**Answer all 3 questions in this part. Each correct answer will receive 4 credits. Clearly indicate the necessary steps, including appropriate formula substitutions, diagrams, graphs, charts, etc. For all questions in this part, a correct numerical answer with no work shown will receive only 1 credit.**   [12 credits]

32 A barrel of fuel oil is a right circular cylinder where the inside measurements of the barrel are a diameter of 22.5 inches and a height of 33.5 inches. There are 231 cubic inches in a liquid gallon. Determine and state, to the *nearest tenth*, the gallons of fuel that are in a barrel of fuel oil.

**33** Given: Parallelogram $ABCD$, $\overline{EFG}$, and diagonal $\overline{DFB}$

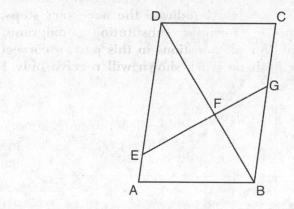

Prove: $\triangle DEF \sim \triangle BGF$

**34** In the diagram below, $\triangle A'B'C'$ is the image of $\triangle ABC$ after a transformation.

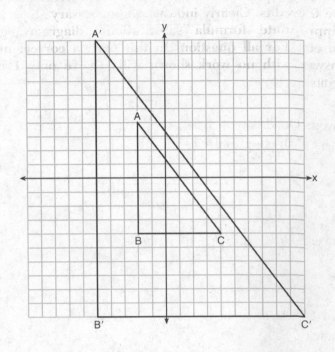

Describe the transformation that was performed.

Explain why $\triangle A'B'C' \sim \triangle ABC$.

## PART IV

Answer the 2 questions in this part. Each correct answer will receive 6 credits. Clearly indicate the necessary steps, including appropriate formula substitutions, diagrams, graphs, charts, etc. For all questions in this part, a correct numerical answer with no work shown will receive only 1 credit. [12 credits]

**35** Given: Quadrilateral $ABCD$ with diagonals $\overline{AC}$ and $\overline{BD}$ that bisect each other, and $\angle 1 \cong \angle 2$

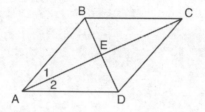

Prove: $\triangle ACD$ is an isosceles triangle and $\triangle AEB$ is a right triangle

**36** A water glass can be modeled by a truncated right cone (a cone which is cut parallel to its base) as shown below.

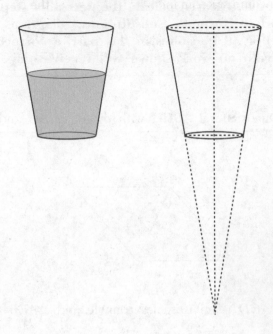

The diameter of the top of the glass is 3 inches, the diameter at the bottom of the glass is 2 inches, and the height of the glass is 5 inches.

The base with a diameter of 2 inches must be parallel to the base with a diameter of 3 inches in order to find the height of the cone. Explain why.

**Question 36 is continued on the next page.**

**Question 36 continued**

Determine and state, in inches, the height of the larger cone.

Determine and state, to the *nearest tenth of a cubic inch*, the volume of the water glass.

# Answers
# June 2016
## Geometry (Common Core)

## Answer Key

### PART I

| | | | | | |
|---|---|---|---|---|---|
| **1.** (3) | **5.** (3) | **9.** (1) | **13.** (2) | **17.** (1) | **21.** (2) |
| **2.** (4) | **6.** (4) | **10.** (2) | **14.** (4) | **18.** (2) | **22.** (3) |
| **3.** (2) | **7.** (3) | **11.** (4) | **15.** (4) | **19.** (2) | **23.** (1) |
| **4.** (1) | **8.** (4) | **12.** (1) | **16.** (3) | **20.** (1) | **24.** (3) |

### PART II

**25.** A reflection over the $x$-axis followed by a translation 6 units right, or a translation 6 units right followed by a reflection over the $x$-axis.

**26.** The coordinates of $P$ are $(12, 2)$.

**27.** $\dfrac{EA}{AC} = \dfrac{EB}{BD}$, therefore $\overline{EC} \parallel \overline{ED}$.

**28.** If a sine equals a cosine, then the angles are complementary.
$\sin(x) = \cos(90° - x)$.
$R = 90° - x$
$R = 90° - 73°$
$R = 17°$

**29.** Using $S = R\theta$ both central angles measure $\dfrac{\pi}{4}$ radians. Dominic is correct.

**30.** The ladder makes a 68° angle with the ground.

**31.** A tangent at point $A$ will be perpendicular to radius $\overline{OA}$. Construct a line perpendicular to the tangent at point $A$.

### PART III

**32.** $V = (13319.86197 \text{ in}^3)\left( \dfrac{1 \text{ gallon}}{231 \text{ in}^3} \right)$
$= 57.7$ gallons

**33.** Use the AA postulate. Parallel sides $\overline{AD}$ and $\overline{BC}$ are cut by transversal $\overline{EG}$ to form congruent alternate interior angles. The second pair of congruent angles are the vertical angles.

**34.** $\angle A \cong \angle A'$, $\angle B \cong \angle B'$, and $\angle C \cong \angle C'$, which makes the two triangles similar by the AA postulate.

### PART IV

**35.** The rhombus properties, consecutive congruent sides and $\perp$ diagonals, can be used to prove $\triangle ABC$ is isosceles and $\triangle AEB$ is a right triangle.

**36.** Volume of the glass
$= V_{\text{large cone}} - V_{\text{small cone}}$
$= 35.3429 \text{ in}^3 - 10.4719 \text{ in}^3$
$= 24.871 \text{ in}^3$
$= 24.9 \text{ in}^3$

In **PARTS II–IV** you are required to show how you arrived at your answers. For sample methods of solutions, see the *Answers Explained* section.

# Answers Explained

## PART I

1. Folding a rectangle in half results in another rectangle, and rotating that rectangle about one of its sides generates a cylinder as shown in the figure.

   The correct choice is **(3)**.

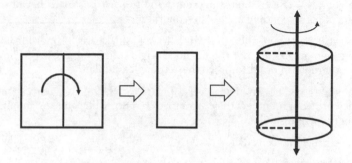

2. A dilation always multiplies lengths by the scale factor, regardless of where the center is located. The length of the 3-inch segment is multiplied by the scale factor 6, so the resulting length is 18 inches.

   The correct choice is **(4)**.

3. Kevin is applying the procedure for completing the square to rewrite the equation in center-radius form, $(x - h)^2 + (y - k)^2 = r^2$, where $h$ and $k$ are the coordinates of the center and $r$ is the radius of the circle. In step 1, Kevin correctly distributes the negative sign on the right side of the equation.

   In step 2, he needs to add $\left(\frac{1}{2}b\right)^2$ to both sides of the equation, where $b$

   is the coefficient of the $x$-term. Since $b = 4$ in this case, $\left(\frac{1}{2}b\right)^2 = \left(\frac{1}{2}4\right)^2 = 4$.

   The error occurs when he added 4 to the left side but subtracted 4 from the right side. He should have added 4 to both sides.

   The correct answer is **(2)**.

4. A translation always preserves slope and parallelism.

   The correct choice is (**1**).

5. Check each choice to see if one of the similarity postulates, AA, SAS, or SSS, does not apply.

   Choice (1)—The triangles can be proven similar by the SAS postulate. Two pairs of corresponding sides are proportional, $\frac{9}{3} = \frac{12}{4} = 3$, and the included angles are congruent.

   Choice (2)—The triangles are similar by the AA postulate because two pairs of corresponding angles are congruent.

   Choice (3)—Even though two pairs of sides are proportional, the congruent angles are not the included angles. $\angle H$ and $\angle K$ would have to be congruent in order to prove those triangles similar.

   Choice (4)—The triangles are similar by SAS. The corresponding sides are congruent, so they are in a 1:1 ratio, which makes them both congruent and similar.

   The correct choice is (**3**).

6. Find the volume of the cube, and subtract the volume of the cone. The radius of the cone is half the diameter, or 4 cm. The side length of the cube is 8.5.

$$V_{\text{cone}} = \frac{1}{3} \pi r^2 h = \frac{1}{3} \pi (4)^2 (8) \qquad V_{\text{cube}} = s^3 = (8.5)^3$$

   The volume of the remaining wood is equal to the difference of the two volumes.

$$V_{\text{remaining wood}} = V_{\text{cube}} - V_{\text{cone}}$$

$$= (8.5)^3 - \frac{1}{3} \pi (4)^2 (8)$$

   The correct choice is (**4**).

7. The SAS postulate states that two right triangles are congruent if two pairs of corresponding sides and the included angles are congruent. In choice (3), the congruent right angles are the included angles formed by the two legs. Therefore, the SAS postulate applies.

The correct choice is **(3)**.

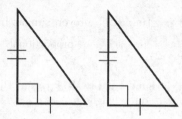

8. One way to eliminate choices when choosing a sequence of transformation is to look at whether the vertices are ordered clockwise or counterclockwise in the image and pre-image. A reflection will always change this direction. The vertices run counterclockwise when going from $A$ to $B$ to $C$ in both $\triangle ABC$ and $\triangle A'B'C'$; therefore, no reflection was used, and choices (1) and (2) can be eliminated. A translation simply slides a shape, so it would not be able to align $\overline{AB}$ with $\overline{A'B'}$; we can eliminate choice (3).

The correct choice is **(4)**.

9. A rhombus has all the properties of a parallelogram plus the following special properties:

- consecutive sides are congruent
- the diagonals are perpendicular
- the diagonals bisect the angles of the rhombus

Any one of these is sufficient to prove a rhombus is a parallelogram.

Choice (1) represents congruent diagonals, which is not necessarily a rhombus property.

Choice (2) demonstrates consecutive congruent sides.

Choice (3) demonstrates perpendicular diagonals.

Choice (4) demonstrates a diagonal bisecting an angle.

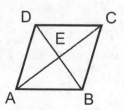

The correct choice is **(1)**.

**10.** $m\angle BOC = m\overset{\frown}{BC}$ because a central angle has the same angle measure as its intercepted arc. $m\angle BAC = \frac{1}{2}m\overset{\frown}{BC}$ because an inscribed angle measures $\frac{1}{2}$ the measure of its intercepted arc. Therefore, $m\angle BAC = \frac{1}{2}m\angle BOC$.

The correct choice is **(2)**.

**11.** The support post, ground, and wall form a right triangle. Use a trigonometric ratio to find the height, $h$.

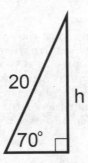

Relative to the 70° angle, $h$ is the opposite and the 20 foot support post is the hypotenuse. This suggests using the sine ratio.

$$\sin(70°) = \frac{opposite}{hypotenuse}$$

$$\sin(70) = \frac{h}{20}$$

$$h = 20\sin(70°)$$
$$= 18.79$$
$$= 18.8$$

The correct choice is **(4)**.

**12.** The slope of the perpendicular bisector of $\overline{NY}$ is the negative reciprocal of the slope of $\overline{NY}$.

$$\text{slope} = \frac{y_2 - y_1}{x_2 - x_1}$$

$$\text{slope of } \overline{NY} = \frac{-7-5}{5-(-11)}$$

$$= \frac{-12}{16}$$

$$= -\frac{3}{4}$$

The slope of the perpendicular bisector $-\frac{4}{3}$

The perpendicular bisector must also pass through the midpoint of $\overline{NY}$.

The midpoint has coordinates $\left(\dfrac{x_1+x_2}{x}, \dfrac{y_1+y_2}{x}\right)$

$$\left(\frac{-11+5}{2}, \frac{5-7}{2}\right)$$

$$(-3, -1)$$

Substitute the coordinates of the midpoint and the slope into the point-slope form of the equation of a line: $y - y_1 = m(x - x_1)$.

$$y - (-1) = \frac{4}{3}(x - (-3))$$

$$y + 1 = \frac{4}{3}(x + 3)$$

The correct choice is **(1)**.

**13.** The altitude to the hypotenuse of a right triangle is the mean proportional to the two parts of the hypotenuse. Use the resulting proportion to solve for $h$, and simplify the resulting radical expression.

$$\frac{RU}{h} = \frac{h}{UT}$$

$$\frac{42-12}{h} = \frac{h}{12}$$

$$h^2 = 12(30)$$

$$h^2 = 360$$

$$h = \sqrt{360}$$

$$= \sqrt{36 \cdot 10}$$

$$= 6\sqrt{10}$$

The correct choice is **(2)**.

**14.** An altitude of a triangle is a segment from a vertex perpendicular to the opposite side. Start by finding the slope of $\overline{BC}$. Using coordinates $B(2, 1)$ and $C(7, 3)$,

$$\text{slope of } \overline{BC} = \frac{y_2 - y_1}{x_2 - x_1}$$

$$= \frac{3-1}{7-2}$$

$$= \frac{2}{5}$$

The slopes of perpendicular lines are negative reciprocals, so the slope of the altitude is $-\frac{5}{2}$.

The correct choice is **(4)**.

15. Write out the cosine and tangent ratios for the two triangles using tangent $= \dfrac{opposite}{adjacent}$ and cosine $\dfrac{adjacent}{hypotenuse}$. It may help to sketch the two triangles separately as well.

$\triangle ERM$          $\triangle JTM$

$\cos(E) = \dfrac{ME}{RE}$       $\cos(T) = \dfrac{TM}{JT}$

$\cos(R) = \dfrac{RM}{RE}$       $\cos(J) = \dfrac{JM}{JT}$

$\tan(E) = \dfrac{RM}{EM}$       $\tan(T) = \dfrac{JM}{TM}$

$\tan(R) = \dfrac{EM}{RM}$       $\tan(J) = \dfrac{TM}{JM}$

Since none of the choices match, the next step is to substitute congruent angles using the similarity statement. From $\triangle ERM \sim \triangle JTM$, we can conclude $\angle E \approx \angle J$ and $\angle R \approx \angle T$. Substituting $\angle E$ for $\angle J$ in $\tan(J) = \dfrac{TM}{JM}$ results in $\tan(E) = \dfrac{TM}{JM}$.

The correct choice is **(4)**.

16. The points undergo the following change of coordinates:

$$A(7, 1) \rightarrow K(7, -1) \quad B(1, 5) \rightarrow L(1, -5)$$
$$C(3, 1) \rightarrow M(3, -1) \quad D(-5, -3) \rightarrow N(-5, 3)$$

In each case the $x$-coordinate remains the same, but the $y$-coordinate changes sign. This is reflection over either the $x$-axis, which flips the point over the axis without changing the distance from the axis.

The correct choice is **(3)**.

**17.** Start by showing that the two triangles are similar. In the figure, $\angle ACD$ $\approx \angle FCB$ because they are vertical angles. $\angle FBC$ and $\angle EBC$ are a linear pair and must sum to 180°.

$$m\angle FBC + \angle EBC = 180°$$

$$m\angle FBC + 115° = 180°$$

$$m\angle FBC = 65°$$

$\angle FBC \approx \angle ADC$ because they have the same measure. $\triangle ACD \sim \triangle FCB$ by the AA postulate, and their corresponding sides are proportional. Set up the appropriate proportion using the similarity statement and solve for the missing side by cross-multiplying.

$$\frac{AC}{FC} = \frac{CD}{CB}$$

$$\frac{6}{15} = \frac{4}{x}$$

$$6x = 60$$

$$x = 10$$

The correct choice is (**1**).

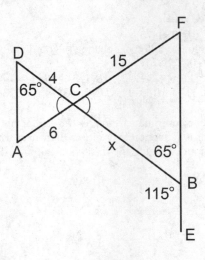

**18.** The relationship between mass, volume, and density is $density = \frac{mass}{volume}$. The mass is given in pounds, and the desired volume is in gallons, so we need to convert the density units from $\frac{ounces}{liter}$ to $\frac{pounds}{gallon}$.

From the reference table we see that 1 gallon = 3.785 liters and 1 pound = 16 ounces. To convert $1.2 \frac{ounces}{liter}$ to $\frac{pounds}{gallon}$, multiply by the conversion ratio for each unit making sure to arrange each ratio so that both the ounces and liters cancel out.

$$density = 1.2 \frac{ounces}{liter} \cdot \frac{1 \ pound}{16 \ ounces} \cdot \frac{3.785 \ liters}{gallon}$$

$$density = (1.2) \frac{3.785 \ pounds}{16 \ gallon}$$

$$density = 0.283875 \frac{pounds}{gallon}$$

Now use the density and a mass of 1 pound to find the volume.

$$density = \frac{mass}{volume}$$

$$0.283875 = \frac{1}{x}$$

$$0.283875x = 1$$

$$x = 3.522 \ pounds$$

$$= 3.5 \ pounds$$

The correct choice is **(2)**.

**19.** Using the given information that $\overline{EA}$ is a perpendicular bisector, we can know

- $\angle ZAE \approx \angle TAE$ because they are both right angles
- $\overline{ZA} \cong \overline{AT}$ because A is the midpoint of $\overline{ZT}$
- $\overline{EA} \cong \overline{EA}$ by the reflexive property

We can conclude that $\triangle ZAT \approx \triangle TAZ$ by the SAS postulate and that all pairs of corresponding sides and angles must be congruent by CPCTC. Now examine each choice.

Choice (1)—$\angle ZEA \approx \angle TEA$ by CPCTC: therefore, $\overline{EA}$ is an angle bisector.

Choice (2)—We know $\overline{EZ} \cong \overline{ET}$, but we do not have enough information to prove $\overline{ZT}$ is congruent to the other two sides.

Choice (3)—A is a midpoint so $\overline{EA}$ must be a median.

Choice (4)—$\angle Z \approx \angle T$ by CPCTC.

The correct choice is **(2)**.

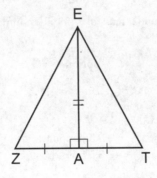

**20.** First, calculate the volume of the hemisphere; then apply the relationship $density = \dfrac{mass}{volume}$ to find the weight. The formula for the volume of a sphere is on the reference table, and we need to divide it by 2 because the tank is only half a sphere. Also, divide the diameter by 2 to get the radius.

$$r = \frac{1}{2} \text{ diameter}$$

$$= \frac{1}{2}(10)$$

$$= 5$$

$$V_{\text{hemisphere}} = \frac{1}{2} \cdot \frac{4}{3}\pi r^3$$

$$= \frac{4}{6}\pi(5)^3$$

$$= 261.799387$$

$$density = \frac{mass}{volume}$$

$$62.4 = \frac{mass}{523.598775}$$

$$mass = 261.799387(62.4)$$

$$= 16{,}336$$

The correct choice is **(1)**.

**21.** A segment drawn inside a triangle parallel to one side will form two similar triangles. Sketching the two triangles separately helps identify the parts needed to write a similarity statement and set up a correct proportion. $\angle C \approx \angle BED$ and $\angle A \approx \angle EDB$ because they are alternate interior angles formed by parallel sides $\overline{AC}$ and $\overline{DE}$, so the triangles are similar by the AA postulate. The similarity statement is $\triangle BCA \sim \triangle BED$.

A proportion formed from corresponding sides will let us find $AC$. The length $AB$ is found by adding the lengths $AD$ and $DE$.

$$\frac{AC}{DE} = \frac{AB}{BD}$$

$$\frac{AC}{4} = \frac{24+12}{12}$$

$$12AC = 4(36)$$

$$12AC = 144$$

$$AC = 12$$

The correct choice is (**2**).

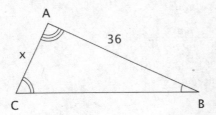

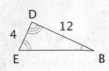

**22.** The area of a triangle equals $\frac{1}{2} base \cdot height$. We can confirm $RS$ is the height to base $ST$ by calculating the slope of each. If the slopes are negative reciprocals, then the segments are perpendicular, and $RS$ is the height. Using coordinates $R(-2, 7)$, $S(-5, 1)$, and $T(7, -5)$, we calculate the slopes.

$$\text{slope} = \frac{y_2 - y_1}{x_2 - x_1}$$

$$\text{slope of } \overline{RS} = \frac{1-7}{-5-(-2)} = \frac{-6}{-3} = \frac{2}{1} \qquad \text{slope of } \overline{ST} = \frac{-5-1}{7-(-5)} = \frac{-6}{12} \ -\frac{1}{2}$$

The slopes are negative reciprocals so $\overline{RS}$ is perpendicular to $\overline{ST}$.

The next step is to calculate the lengths $RS$ and $ST$ using the distance formula.

$$RS = \sqrt{\left(x_1 - x_2\right)^2 + \left(y_1 - y_2\right)^2} \qquad ST = \sqrt{\left(x_1 - x_2\right)^2 + \left(y_1 - y_2\right)^2}$$

$$= \sqrt{\left(-2-(-5)\right)^2 + (7-1)^2} \qquad = \sqrt{(-5-7)^2 + \left(1-(-5)\right)^2}$$

$$= \sqrt{3^2 + 6^2} \qquad\qquad\qquad = \sqrt{(-12)^2 + 6^2}$$

$$= \sqrt{45} \qquad\qquad\qquad\qquad = \sqrt{180}$$

Now substitute these values into the area formula.

$$\text{Area} = \frac{1}{2} base \cdot height$$

$$= \frac{1}{2} \sqrt{45}\sqrt{180}$$

$$= \frac{1}{2} \sqrt{8100}$$

$$= \frac{1}{2} (90)$$

$$= 45$$

The correct choice is **(3)**.

**23.** In a circle, the perpendicular bisector of any chord passes through the center. We can use this theorem to locate the center of the circle. Sketch $\overrightarrow{PQ}$, the perpendicular bisector of $\overline{AB}$, as shown. The center of the circle lies along this line 2 units from $\overline{AB}$. The two possible locations of the center are points $P(1, -2)$ and $Q(5, -2)$.

The equation of a circle is $(x - h)^2 + (y - k)^2 = R^2$, where $h$ and $k$ are the coordinates of the center of the circle. Given the two possible locations of the center, either $h = 1$ and $k = -2$, or $h = 5$ and $k = -2$. The only choices that match these values are $(x - 1)^2 + (y + 2)^2 = 29$ and $(x - 5)^2 + (y + 2)^2 = 25$.

We now need to find the radius of the circle. The Pythagorean theorem can be used in $\triangle POS$ to find $PA^2$. Using $PO = 2$ and $OA = 5$, we find

$$PO^2 + OA^2 = PA^2$$
$$2^2 + 5^2 = PA^2$$
$$PA^2 = 29$$

The square of the radius is 29, so the only choice with the correct center and radius is $(x - 1)^2 + (y + 2)^2 = 29$.

The correct choice is (**1**).

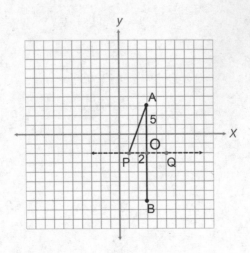

**24.** Use the formula for the area of a sector, where $r$ is the radius and $\theta$ is the central angle measured in degrees.

$$A_{\text{sector}} = \frac{\theta}{360°}\pi R^2$$

$$= \frac{60°}{360°}\pi(8)^2$$

$$= \frac{1}{6}\pi(64)$$

$$= \frac{32\pi}{3}$$

The correct choice is **(3)**.

## PART II

**25.** One clue to use when trying to identify a transformation is to look at whether the vertices are ordered clockwise or counterclockwise in the image and pre-image. A reflection will always change this direction. The vertices run counterclockwise when going from *A* to *B* to *C* in △*ABC* and counterclockwise in △*DEF*; therefore, a reflection was used. The first transformation is a reflection over the *x*-axis to map △*ABC* to △*A'B'C'* shown in the figure. From there, a translation 6 units right maps △*A'B'C'* to △*DEF*. For these two transformations, the order does not matter.

The correct answer is a reflection over the *x*-axis followed by a translation 6 units right, or a translation 6 units right followed by a reflection over the *x*-axis.

An alternate solution is a rotation of 180° about the point (4, 0) followed by a reflection over the line *x* = 7.

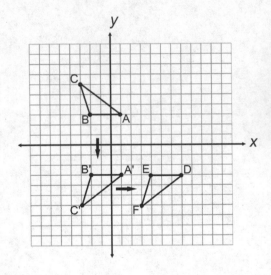

**26.** Find the $x$- and $y$-coordinates of $P$ with the formula

$$ratio = \frac{x - x_1}{x_2 - x}, ratio = \frac{y - y_1}{y_2 - y}$$

Using the coordinates $A(4, 2)$ and $B(22, 2)$, we have $x_1 = 4$, $y_1 = 2$, $x_2 = 22$, $y_2 = 2$.

The ratio is $\frac{4}{5}$.

Solve for the $x$-coordinate:

$$ratio = \frac{x - x_1}{x_2 - x}$$

$$\frac{4}{5} = \frac{x - 4}{22 - x}$$

$$5(x - 4) = 4(22 - x) \qquad \text{cross-multiply}$$
$$5x - 20 = 88 - 4x$$
$$9x = 108$$
$$x = 12$$

Solve for the $y$-coordinate:

$$ratio = \frac{y - y_1}{y_2 - y}$$

$$\frac{4}{5} = \frac{y - 2}{2 - y}$$

$$5(y - 2) = 4(2 - y) \qquad \text{cross-multiply}$$
$$5y - 10 = 8 - 4y$$
$$9y = 18$$
$$y = 2$$

The coordinates of $P$ are $(12, 2)$.

**27.** If a segment is drawn inside a triangle parallel to one side, then it divides the sides proportionally. Using this theorem in the opposite direction, we can prove $\overline{AB}$ // $\overline{CD}$ by showing sides $\overline{EC}$ and $\overline{ED}$ are divided proportionally.

$$\frac{EA}{AC} = \frac{EB}{BD}$$

$$\frac{3.75}{5} = \frac{4.5}{6}$$

$$6(3.75) = 5(4.5) \quad \text{cross-multiply}$$

$$22.5 = 22.5$$

The proportional is valid; therefore, $\overline{EC}$ // $\overline{ED}$.

An alternative approach would have been to prove $\triangle EAB$ is similar to to $\triangle ECD$. You can show that $\frac{EA}{AC} = \frac{EB}{ED}$, and that the included angle $\angle E$ is a shared angle so it is congruent to itself by the reflexive property.

The triangles are similar by the SAS postulate. $\angle EAB$ would then be congruent to $\angle ECD$ because corresponding angles in similar triangles are congruent, and $\overline{EC}$ would be parallel to $\overline{ED}$ because alternate interior angles are congruent.

**28.** The co-function relationship for the sine and cosine ratios states

$$\sin(x) = \cos(90° - x)$$

We are given $\sin(73°) = \cos(R)$; therefore,

$$R = 90° - x$$

$$R = 90° - 73°$$

$$R = 17°$$

Be sure to provide an *explanation* for full credit. If a sine is equal to a cosine, the angles are complementary.

**29.** The length of an arc is given by the formula

$$S = R\theta$$

where $S$ is the arc length, $R$ is the radius of the circle, and $\theta$ is the measure of the central angle in radians. Use this formula to find the two central angles.

| Circle 1 | Circle 2 |
|---|---|
| $S = R\theta$ | $S = R\theta$ |
| $\pi = (4)A$ | $\dfrac{13\pi}{8} = (6.5)B$ |
| $A = \dfrac{\pi}{4}$ radians | $B = \dfrac{\frac{13\pi}{8}}{6.5}$ |
| | $= \dfrac{13\pi}{8}$ |
| | $= \dfrac{\pi}{4}$ radians |

Be sure to provide a full explanation. Dominic is correct because central angles $A$ and $B$ were calculated using $S = R\theta$ and they have the same measure.

**30.** The ladder, building, and ground form a right triangle that is shown in the figure. The unknown quantity is an angle, so an inverse trigonometric function is needed. Relative to the angle $x$, the 10-foot dimension is the opposite, and the 4-foot dimension is the adjacent. The tangent ratio applies here.

$$\text{tangent} = \frac{opposite}{adjacent}$$

$$\tan(x) = \frac{10}{4}$$

$$x = \tan^{-1}\left(\frac{10}{4}\right) \quad \text{apply the inverse tangent to find the angle}$$

$$= 68.1985°$$

$$= 68°$$

The ladder makes a 68° angle with the ground.

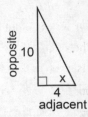

**31.** A tangent at point *A* will be perpendicular to radius $\overline{OA}$, so we need to construct a line perpendicular to the tangent at point *A*.

**STEP 1:** Extend $\overline{OA}$ to point *B* so that *AB* is longer than *OA*.

**STEP 2:** With the point of the compass at *A*, make two arcs intersecting $\overline{OB}$ at *C* and *D*.

**STEP 3:** With the point of the compass at *C*, make a pair of arcs on each side of $\overline{OB}$.

**STEP 4:** Repeat with the point of the compass at *D*. The arcs will intersect at *E* and *F*.

**STEP 5:** Connect points *E*, *A*, and *F*. $\overline{EAF}$ is a tangent to point *A*.

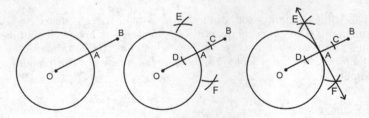

# PART III

**32.** First use the volume formula for a cylinder to calculate the volume of the barrel in cubic inches. The radius of the barrel is $\frac{1}{2}$ the diameter, or 11.25 inches, and the height is given as 33.5 inches.

$$V = \pi r^2 h$$
$$= \pi (11.25)^2 (33.5)$$
$$= 13{,}319.86197 \text{ in}^3$$

Next convert the volume to gallons using the conversion ratio. Note that we want to keep gallons and eliminate cubic inches, so we will write the ratio as $\dfrac{1 \text{ gallon}}{231 \text{ in}^3}$.

$$V = (13{,}319.86197 \text{ in}^3)\left( \frac{1 \text{ gallon}}{231 \text{ in}^3} \right)$$
$$= 56.6617 \text{ gallons}$$
$$= 57.7 \text{ gallons}$$

**33.** $\triangle DEF$ and $\triangle BGF$ can be proven similar using the AA postulate. Parallel sides $\overline{AD}$ and $\overline{BC}$ are cut by transversal $\overline{EG}$ to form congruent alternate interior angles. The second pair of congruent angles are the vertical angles. A two-column proof is shown below.

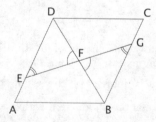

| Statement | Reason |
|---|---|
| 1. Parallelogram $ABCD$, $\overline{EFG}$, and diagonal $\overline{DFB}$ | 1. Given |
| 2. $\overline{AD} \parallel \overline{BC}$ | 2. Opposite sides of a parallelogram are parallel. |
| 3. $\angle DEF \approx \angle BGF$ | 3. Alternate interior angles formed by parallel lines are congruent. |
| 4. $\angle EFD \approx \angle BFG$ | 4. Intersecting lines form congruent vertical angles. |
| 5. $\triangle DEF \sim \triangle BGF$ | 5. AA |

**34.** A dilation was clearly performed on $\triangle ABC$. To fully specify the dilation, we need to find the scale factor and center of dilation. The scale factor can be calculated by finding the ratio of a pair of corresponding sides. Sides $\overline{BC}$ and $\overline{B'C'}$ are a convenient pair. The length of $\overline{BC}$ is 6 units. The length of $\overline{B'C'}$ is 15 units.

$$\text{Scale factor} = \frac{B'C'}{BC}$$

$$= \frac{15}{6}$$

$$= \frac{5}{2}$$

One method to identify the center of dilation is to sketch segments between corresponding points. The point of intersection is the center of dilation. Lines $\overleftrightarrow{AA'}$, $\overleftrightarrow{BB'}$ intersect at the origin, so the center of dilation is $(0, 0)$. The transformation is a dilation with a scale factor of $\frac{5}{2}$ about the origin.

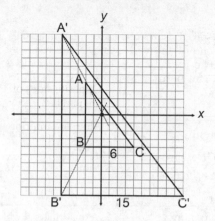

There are a number of ways to explain why $\triangle A'B'C'$ is similar to $\triangle ABC$. One possible answer is that a dilation preserves angle measure. $\angle A \approx \angle A'$, $\angle B \approx \angle B'$, and $\angle C \approx \angle C'$, which makes the two triangles similar by the AA postulate.

Another possible explanation is that the dilation multiplies each side length by the same scale factor, so all three pairs of corresponding sides are proportional and the triangles are similar by SSS.

## PART IV

**35.** The key to this proof is to recognize that diagonals that bisect both each other and one of the vertex angles are sufficient to prove *ABCD* is a rhombus. The rhombus properties, consecutive congruent sides and ⊥ diagonals, can be used to prove △*ABC* is isosceles and △*AEB* is a right triangle.

| Statement | Reason |
|---|---|
| 1. Quadrilateral *ABCD*, diagonals $\overline{AC}$ and $\overline{BD}$ bisect each other, ∠1 ≈ ∠2 | 1. Given |
| 2. *ABCD* is a parallelogram | 2. A quadrilateral whose diagonals bisect each other is a parallelogram |
| 3. *ABCD* is a rhombus | 3. A parallelogram whose diagonal bisects vertex angles is a rhombus |
| 4. $\overline{AD} \approx \overline{CD}$ | 4. Consecutive sides of a rhombus are congruent |
| 5. △*ABC* is isosceles | 5. A triangle with two congruent sides is isosceles |
| 6. $\overline{AC} \perp \overline{BD}$ | 6. The diagonals of a rhombus are perpendicular |
| 7. ∠*AEB* is a right angle | 7. Perpendicular lines form right angles |
| 8. △*AEB* is a right triangle | 8. A triangle with a right angle is a right triangle |

**36.** You can think of the figure as a small cone and a large cone. The volume of the water glass is simply the volume of the large cone minus the volume of the small cone.

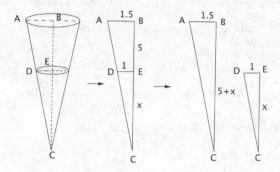

The two bases must be parallel to each other in order to find the heights. If they are parallel, then $\triangle ABC$ is similar to $\triangle DEC$, and a proportion can be used to find $BC$ and $EC$.

Sketching the triangles separately as shown will help determine the appropriate expressions for each side. $AB$ and $DE$ are radii of the bases, so each is equal to half the diameter. Letting $EC$ equal $x$, $BC = 5 + x$. We can now write a proportion using corresponding sides and cross-multiply to solve for $x$.

$$\frac{AB}{DE} = \frac{BC}{EC}$$

$$\frac{1.5}{1} = \frac{x+5}{x}$$

$$1.5x = 1(x + 5)$$

$$1.5x = x + 5$$

$$0.5x = 5$$

$$x = 10$$

$$BC = x + 5 = 15$$

The height of the larger cone is 15 inches.

Finally, find the volume of each cone using $V = \frac{1}{3}\pi r^2 h$ and subtract to find the volume of the water glass.

Large cone

$V = \frac{1}{3}\pi r^2 h$

$= \frac{1}{3}\pi(1.5)^2(15)$

$= 35.3429 \text{ in}^3$

Small cone

$V = \frac{1}{3}\pi r^2 h$

$= \frac{1}{3}\pi(1)^2(10)$

$= 10.4719 \text{ in}^3$

Volume of the glass $= V_{\text{large cone}} - V_{\text{small cone}}$

$= 35.3429 \text{ in}^3 - 10.4719 \text{ in}^3$

$= 24.871 \text{ in}^3$

$= 24.9 \text{ in}^3$

| Topic | Question Numbers | Number of Points | Your Points | Your Percentage |
|---|---|---|---|---|
| 1. Basic Angle and Segment Relationships | | | | |
| 2. Angle and Segment Relationships in Triangles and Polygons | 13, 19 | 2 + 2 = 4 | | |
| 3. Constructions | 31 | 2 | | |
| 4. Transformations | 4, 8, 16, 25, 34 | 2 + 2 + 2 + 2 + 4 = 12 | | |
| 5. Triangle Congruence | 7, 35 | 2 + 6 = 8 | | |
| 6. Lines, Segments and Circles on the Coordinate Plane | 3, 6, 12, 14, 22, 23, 26 | 2 + 2 + 2 + 2 + 2 + 2 + 2 = 14 | | |
| 7. Similarity | 2, 5, 17, 21, 27 | 2 + 2 + 2 + 2 + 2 = 10 | | |
| 8. Trigonometry | 11, 15, 28, 30 | 2 + 2 + 2 + 2 = 8 | | |
| 9. Parallelograms | 9, 33 | 2 + 4 = 6 | | |
| 10. Coordinate Geometry Proofs | | | | |
| 11. Volume and Solids | 1, 20, 36 | 2 + 2 + 6 = 10 | | |
| 12. Modeling | 18, 32 | 2 + 4 = 6 | | |
| 13. Circles | 10, 24, 29 | 2 + 2 + 2 = 6 | | |

# HOW TO CONVERT YOUR RAW SCORE TO YOUR GEOMETRY REGENTS EXAMINATION SCORE

The conversion chart below can be used to determine your final score on the June 2016 Regents Exam in Geometry. To find your final exam score, locate in the "Raw Score" column the total number of points you scored out of a possible 86. Then locate in the adjacent column to the right the scale score that corresponds to your raw score. The scale score is your final score.

| Raw Score | Scale Score | Performance Level | Raw Score | Scale Score | Performance Level | Raw Score | Scale Score | Performance Level |
|---|---|---|---|---|---|---|---|---|
| 86 | 100 | 5 | 57 | 79 | 3 | 28 | 59 | 2 |
| 85 | 99 | 5 | 56 | 78 | 3 | 27 | 58 | 2 |
| 84 | 98 | 5 | 55 | 78 | 3 | 26 | 57 | 2 |
| 83 | 97 | 5 | 54 | 77 | 3 | 25 | 55 | 2 |
| 82 | 96 | 5 | 53 | 77 | 3 | 24 | 54 | 1 |
| 81 | 95 | 5 | 52 | 76 | 3 | 23 | 53 | 1 |
| 80 | 94 | 5 | 51 | 76 | 3 | 22 | 51 | 1 |
| 79 | 93 | 5 | 50 | 75 | 3 | 21 | 50 | 1 |
| 78 | 92 | 5 | 49 | 75 | 3 | 20 | 48 | 1 |
| 77 | 91 | 5 | 48 | 74 | 3 | 19 | 47 | 1 |
| 76 | 90 | 5 | 47 | 74 | 3 | 18 | 45 | 1 |
| 75 | 89 | 5 | 46 | 73 | 3 | 17 | 43 | 1 |
| 74 | 89 | 5 | 45 | 73 | 3 | 16 | 41 | 1 |
| 73 | 88 | 5 | 44 | 72 | 3 | 15 | 39 | 1 |
| 72 | 87 | 5 | 43 | 71 | 3 | 14 | 38 | 1 |
| 71 | 87 | 5 | 42 | 71 | 3 | 13 | 35 | 1 |
| 70 | 86 | 5 | 41 | 70 | 3 | 12 | 33 | 1 |
| 69 | 86 | 5 | 40 | 69 | 3 | 11 | 31 | 1 |
| 68 | 85 | 5 | 39 | 69 | 3 | 10 | 29 | 1 |
| 67 | 84 | 4 | 38 | 68 | 3 | 9 | 27 | 1 |
| 66 | 83 | 4 | 37 | 67 | 3 | 8 | 24 | 1 |
| 65 | 83 | 4 | 36 | 66 | 3 | 7 | 22 | 1 |
| 64 | 82 | 4 | 35 | 66 | 3 | 6 | 19 | 1 |
| 63 | 82 | 4 | 34 | 65 | 3 | 5 | 16 | 1 |
| 62 | 81 | 4 | 33 | 64 | 3 | 4 | 13 | 1 |
| 61 | 81 | 4 | 32 | 63 | 2 | 3 | 10 | 1 |
| 60 | 80 | 4 | 31 | 62 | 2 | 2 | 7 | 1 |
| 59 | 80 | 4 | 30 | 61 | 2 | 1 | 4 | 1 |
| 58 | 79 | 3 | 29 | 60 | 2 | 0 | 0 | 1 |

# Examination
# August 2016
## Geometry (Common Core)

## GEOMETRY REFERENCE SHEET

| | |
|---|---|
| 1 inch = 2.54 centimeters | 1 ton = 2000 pounds |
| 1 meter = 39.37 inches | 1 cup = 8 fluid ounces |
| 1 mile = 5280 feet | 1 pint = 2 cups |
| 1 mile = 1760 yards | 1 quart = 2 pints |
| 1 mile = 1.609 kilometers | 1 gallon = 4 quarts |
| 1 kilometer = 0.62 mile | 1 gallon = 3.785 liters |
| 1 pound = 16 ounces | 1 liter = 0.264 gallon |
| 1 pound = 0.454 kilogram | 1 liter = 1000 cubic centimeters |
| 1 kilogram = 2.2 pounds | |

| | |
|---|---|
| Triangle | $A = \frac{1}{2}bh$ |
| Parallelogram | $A = bh$ |
| Circle | $A = \pi r^2$ |
| Circle | $C = \pi d$ or $C = 2\pi r$ |
| General Prisms | $V = Bh$ |
| Cylinder | $V = \pi r^2 h$ |
| Sphere | $V = \frac{4}{3}\pi r^3$ |

| Cone | $V = \frac{1}{3}\pi r^2 h$ |
|---|---|
| Pyramid | $V = \frac{1}{3}Bh$ |
| Pythagorean Theorem | $a^2 + b^2 = c^2$ |
| Quadratic Formula | $x = \dfrac{-b \pm \sqrt{b^2 - 4ac}}{2a}$ |
| Arithmetic Sequence | $a_n = a_1 + (n-1)d$ |
| Geometric Sequence | $a_n = a_1 r^{n-1}$ |
| Geometric Series | $S_n = \dfrac{a_1 - a_1 r^n}{1-r}$ where $r \neq 1$ |
| Radians | 1 radian $= \dfrac{180}{\pi}$ degrees |
| Degrees | 1 degree $= \dfrac{\pi}{180}$ radians |
| Exponential Growth/Decay | $A = A_0 e^{k(t - t_0)} + B_0$ |

## PART I

Answer all 24 questions in this part. Each correct answer will receive 2 credits. No partial credit will be allowed. For each statement or question, write in the space provided the numeral preceding the word or expression that best completes the statement or answers the question. [48 credits]

**1** In the diagram below, lines $\ell$, $m$, $n$, and $p$ intersect line $r$.

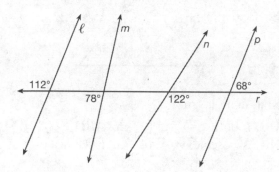

Which statement is true?

(1) $\ell \,/\!/\, n$          (3) $m \,/\!/\, p$

(2) $\ell \,/\!/\, p$          (4) $m \,/\!/\, n$          1 _____

**2** Which transformation would *not* always produce an image that would be congruent to the original figure?

(1) translation          (3) rotation

(2) dilation          (4) reflection          2 _____

**3** If an equilateral triangle is continuously rotated around one of its medians, which 3-dimensional object is generated?

(1) cone            (3) prism
(2) pyramid         (4) sphere            3 _____

**4** In the diagram below, m∠BDC = 100°, m∠A = 50°, and m∠DBC = 30°.

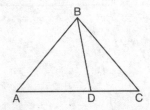

Which statement is true?

(1) △ABD is obtuse.          (3) m∠ABD = 80°
(2) △ABC is isosceles.       (4) △ABD is scalene.        4 _____

**5** Which point shown in the graph below is the image of point *P* after a counterclockwise rotation of 90° about the origin?

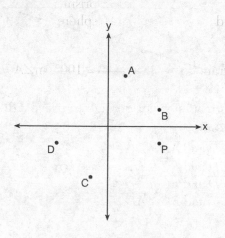

(1) *A*                (3) *C*

(2) *B*                (4) *D*                    5 _____

**6** In △*ABC*, where ∠*C* is a right angle, $\cos A = \dfrac{\sqrt{21}}{5}$. What is sin *B*?

(1) $\dfrac{\sqrt{21}}{5}$        (3) $\dfrac{2}{5}$

(2) $\dfrac{\sqrt{21}}{2}$        (4) $\dfrac{5}{\sqrt{21}}$        6 _____

7  Quadrilateral *ABCD* with diagonals $\overline{AC}$ and $\overline{BD}$ is shown in the diagram below.

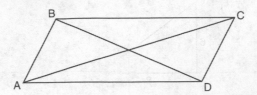

Which information is *not* enough to prove *ABCD* is a parallelogram?

(1) $\overline{AB} \cong \overline{CD}$ and $\overline{AB} \parallel \overline{DC}$
(2) $\overline{AB} \cong \overline{CD}$ and $\overline{BC} \cong \overline{DA}$
(3) $\overline{AB} \cong \overline{CD}$ and $\overline{BC} \parallel \overline{AD}$
(4) $\overline{AB} \parallel \overline{DC}$ and $\overline{BC} \parallel \overline{AD}$

7 _____

8  An equilateral triangle has sides of length 20. To the *nearest tenth*, what is the height of the equilateral triangle?

(1) 10.0            (3) 17.3
(2) 11.5            (4) 23.1

8 _____

**9** Given: $\triangle AEC$, $\triangle DEF$, and $\overline{FE} \perp \overline{CE}$

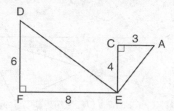

What is a correct sequence of similarity transformations that shows $\triangle AEC \sim \triangle DEF$?

(1) a rotation of 180 degrees about point $E$ followed by a horizontal translation

(2) a counterclockwise rotation of 90 degrees about point $E$ followed by a horizontal translation

(3) a rotation of 180 degrees about point $E$ followed by a dilation with a scale factor of 2 centered at point $E$

(4) a counterclockwise rotation of 90 degrees about point $E$ followed by a dilation with a scale factor of 2 centered at point $E$

9 _____

**10** In the diagram of right triangle $ABC$, $\overline{CD}$ intersects hypotenuse $\overline{AB}$ at $D$.

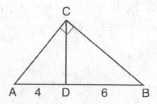

If $AD = 4$ and $DB = 6$, which length of $\overline{AC}$ makes $\overline{CD} \perp \overline{AB}$ ?

(1) $2\sqrt{6}$          (3) $2\sqrt{15}$

(2) $2\sqrt{10}$         (4) $4\sqrt{2}$

10 _____

**11** Segment *CD* is the perpendicular bisector of $\overline{AB}$ at *E*. Which pair of segments does *not* have to be congruent?

(1) $\overline{AD}, \overline{BD}$        (3) $\overline{AE}, \overline{BE}$

(2) $\overline{AC}, \overline{BC}$        (4) $\overline{DE}, \overline{CE}$      11 _____

**12** In triangle *CHR*, *O* is on $\overline{HR}$, and *D* is on $\overline{CR}$ so that $\angle H \cong \angle RDO$.

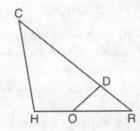

If *RD* = 4 and *RO* = 6, and *OH* = 4, what is the length of $\overline{CD}$ ?

(1) $2\dfrac{2}{3}$        (3) 11

(2) $6\dfrac{2}{3}$        (4) 15      12 _____

**13** The cross section of a regular pyramid contains the altitude of the pyramid. The shape of this cross section is a

(1) circle        (3) triangle

(2) square        (4) rectangle      13 _____

**14** The diagonals of rhombus *TEAM* intersect at $P(2, 1)$. If the equation of the line that contains diagonal $\overline{TA}$ is $y = -x + 3$, what is the equation of a line that contains diagonal $\overline{EM}$?

(1) $y = x - 1$        (3) $y = -x - 1$

(2) $y = x - 3$        (4) $y = -x - 3$      14 _____

**15** The coordinates of vertices *A* and *B* of $\triangle ABC$ are $A(3, 4)$ and $B(3, 12)$. If the area of $\triangle ABC$ is 24 square units, what could be the coordinates of point *C*?

(1) $(3, 6)$        (3) $(-3, 8)$

(2) $(8, -3)$        (4) $(6, 3)$      15 _____

**16** What are the coordinates of the center and the length of the radius of the circle represented by the equation $x^2 + y^2 - 4x + 8y + 11 = 0$?

(1) center $(2, -4)$ and radius 3

(2) center $(-2, 4)$ and radius 3

(3) center $(2, -4)$ and radius 9

(4) center $(-2, 4)$ and radius 9      16 _____

**17** The density of the American white oak tree is 752 kilograms per cubic meter. If the trunk of an American white oak tree has a circumference of 4.5 meters and the height of the trunk is 8 meters, what is the approximate number of kilograms of the trunk?

(1) 13        (3) 13,536

(2) 9694        (4) 30,456      17 _____

18 Point $P$ is on the directed line segment from point $X(-6, -2)$ to point $Y(6, 7)$ and divides the segment in the ratio 1:5. What are the coordinates of point $P$?

(1) $(4, 5\frac{1}{2})$

(3) $(-4\frac{1}{2}, 0)$

(2) $(-\frac{1}{2}, -4)$

(4) $(-4, -\frac{1}{2})$

18 _____

*19 In circle $O$, diameter $\overline{AB}$, chord $\overline{BC}$, and radius $\overline{OC}$ are drawn, and the measure of arc $BC$ is $108°$.

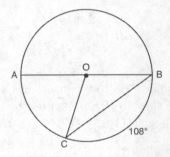

Some students wrote these formulas to find the area of sector $COB$:

Amy     $\frac{3}{10} \cdot \pi \cdot (BC)^2$

Beth     $\frac{108}{360} \cdot \pi \cdot (OC)^2$

Carl     $\frac{3}{10} \cdot \pi \cdot \frac{1}{2}(AB)^2$

Dex     $\frac{108}{360} \cdot \pi \cdot \frac{1}{2}(AB)^2$

Which students wrote correct formulas?

(1) Amy and Dex          (3) Carl and Amy
(2) Beth and Carl         (4) Dex and Beth

19 _____

*Due to a typographical error, question 19 does not have a correct answer.
 All students received credit for question 19.

**20** Tennis balls are sold in cylindrical cans with the balls stacked one on top of the other. A tennis ball has a diameter of 6.7 cm. To the *nearest cubic centimeter*, what is the minimum volume of the can that holds a stack of 4 tennis balls?

(1) 236           (3) 564

(2) 282           (4) 945         20 _____

**21** Line segment $A'B'$, whose endpoints are $(4, -2)$ and $(16, 14)$, is the image of $\overline{AB}$ after a dilation of $\frac{1}{2}$ centered at the origin. What is the length of $\overline{AB}$?

(1) 5           (3) 20

(2) 10          (4) 40         21 _____

**22** Given: $\triangle ABE$ and $\triangle CBD$ shown in the diagram below with $\overline{DB} \cong \overline{BE}$

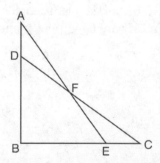

Which statement is needed to prove $\triangle ABE \cong \triangle CBD$ using only SAS $\cong$ SAS?

(1) $\angle CDB \cong \angle AEB$      (3) $\overline{AD} \cong \overline{CE}$

(2) $\angle AFD \cong \angle EFC$      (4) $\overline{AE} \cong \overline{CD}$        22 _____

**23** In the diagram below, $\overline{BC}$ is the diameter of circle $A$.

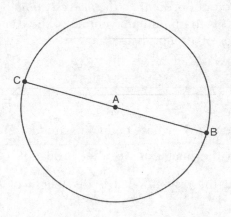

Point $D$, which is unique from points $B$ and $C$, is plotted on circle $A$. Which statement must always be true?

(1) $\triangle BCD$ is a right triangle.

(2) $\triangle BCD$ is an isosceles triangle.

(3) $\triangle BAD$ and $\triangle CBD$ are similar triangles.

(4) $\triangle BAD$ and $\triangle CAD$ are congruent triangles.

23 _____

**24** In the diagram below, $ABCD$ is a parallelogram, $\overline{AB}$ is extended through $B$ to $E$, and $\overline{CE}$ is drawn.

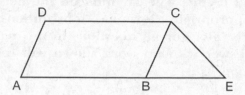

If $\overline{CE} \cong \overline{BE}$ and m$\angle D = 112°$, what is m$\angle E$?

(1) 44°                    (3) 68°
(2) 56°                    (4) 112°                    24 _____

## PART II

Answer all 7 questions in this part. Each correct answer will receive 2 credits. Clearly indicate the necessary steps, including appropriate formula substitutions, diagrams, graphs, charts, etc. For all questions in this part, a correct numerical answer with no work shown will receive only 1 credit.    [14 credits]

**25** Lines *AE* and *BD* are tangent to circles *O* and *P* at *A*, *E*, *B*, and *D*, as shown in the diagram below.

If *AC* : *CE* = 5 : 3, and *BD* = 56, determine and state the length of $\overline{CD}$.

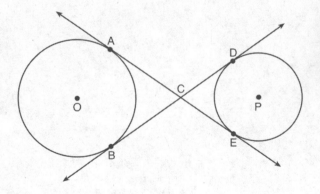

**26** In the diagram below, $\triangle ABC$ has coordinates $A(1, 1)$, $B(4, 1)$, and $C(4, 5)$. Graph and label $\triangle A''B''C''$, the image of $\triangle ABC$ after the translation five units to the right and two units up followed by the reflection over the line $y = 0$.

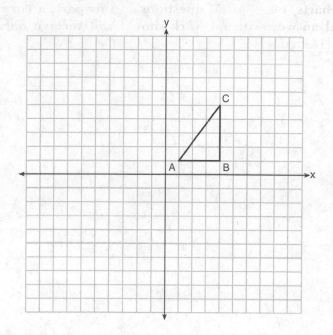

**27** A regular hexagon is rotated in a counterclockwise direction about its center. Determine and state the minimum number of degrees in the rotation such that the hexagon will coincide with itself.

**28** In the diagram of $\triangle ABC$ shown below, use a compass and straightedge to construct the median to $\overline{AB}$. [Leave all construction marks.]

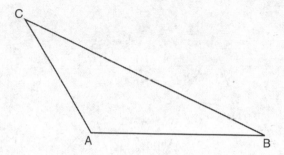

**29** Triangle *MNP* is the image of triangle *JKL* after a 120° counter-clockwise rotation about point *Q*. If the measure of angle *L* is 47° and the measure of angle *N* is 57°, determine the measure of angle *M*. Explain how you arrived at your answer.

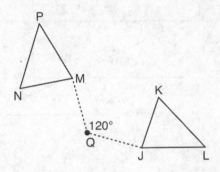

**30** A circle has a center at $(1, -2)$ and radius of 4. Does the point $(3.4, 1.2)$ lie on the circle? Justify your answer.

**31** In the diagram below, a window of a house is 15 feet above the ground. A ladder is placed against the house with its base at an angle of 75° with the ground. Determine and state the length of the ladder to the *nearest tenth of a foot*.

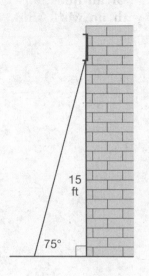

## PART III

Answer all 3 questions in this part. Each correct answer will receive 4 credits. Clearly indicate the necessary steps, including appropriate formula substitutions, diagrams, graphs, charts, etc. For all questions in this part, a correct numerical answer with no work shown will receive only 1 credit.   [12 credits]

32 Using a compass and straightedge, construct and label $\triangle A'B'C'$, the image of $\triangle ABC$ after a dilation with a scale factor of 2 and centered at $B$. [Leave all construction marks.]

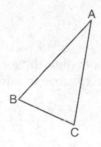

Describe the relationship between the lengths of $\overline{AC}$ and $\overline{A'C'}$.

**33** The grid below shows △ABC and △DEF.

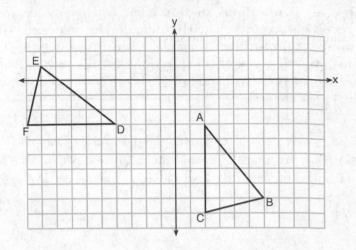

Let △A′B′C′ be the image of △ABC after a rotation about point A. Determine and state the location of B′ if the location of point C′ is (8, –3). Explain your answer.

Is △DEF congruent to △A′B′C′? Explain your answer.

**34** As modeled below, a movie is projected onto a large outdoor screen. The bottom of the 60-foot-tall screen is 12 feet off the ground. The projector sits on the ground at a horizontal distance of 75 feet from the screen.

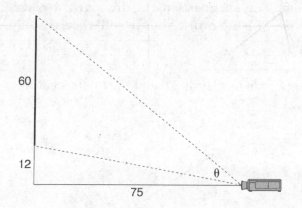

Determine and state, to the *nearest tenth of a degree*, the measure of $\theta$, the projection angle.

## PART IV

Answer the 2 questions in this part. Each correct answer will receive 6 credits. Clearly indicate the necessary steps, including appropriate formula substitutions, diagrams, graphs, charts, etc. For all questions in this part, a correct numerical answer with no work shown will receive only 1 credit. [12 credits]

**35** Given: Circle $O$, chords $\overline{AB}$ and $\overline{CD}$ intersect at $E$

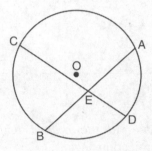

Theorem: If two chords intersect in a circle, the product of the lengths of the segments of one chord is equal to the product of the lengths of the segments of the other chord.

Prove this theorem by proving $AE \cdot EB = CE \cdot ED$.

**36** A snow cone consists of a paper cone completely filled with shaved ice and topped with a hemisphere of shaved ice, as shown in the diagram below. The inside diameter of both the cone and the hemisphere is 8.3 centimeters. The height of the cone is 10.2 centimeters.

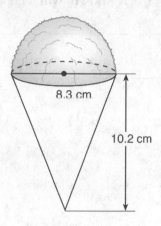

8.3 cm

10.2 cm

The desired density of the shaved ice is 0.697 g/cm$^3$, and the cost, per kilogram, of ice is \$3.83. Determine and state the cost of the ice needed to make 50 snow cones.

# Answers
# August 2016

## Geometry (Common Core)

## Answer Key

### PART I

| | | | | | |
|---|---|---|---|---|---|
| **1.** (2) | **5.** (1) | **9.** (4) | **13.** (3) | **17.** (2) | **21.** (4) |
| **2.** (2) | **6.** (1) | **10.** (2) | **14.** (1) | **18.** (4) | **22.** (3) |
| **3.** (1) | **7.** (3) | **11.** (4) | **15.** (3) | **19.** (*) | **23.** (1) |
| **4.** (2) | **8.** (3) | **12.** (3) | **16.** (1) | **20.** (4) | **24.** (1) |

*Due to a typographical error, question 19 does not have a correct answer.
All students received credit for question 19.

### PART II

**25.** 21

**26.** $A''(6, -3)$, $B''(9, -3)$, $C''(9, -7)$

**27.** $60°$

**28.** Construct point $D$, the midpoint of $\overline{AB}$. $\overline{CD}$ is the median to side $\overline{AB}$.

**29.** $76°$

**30.** Yes, because $(3.4, 1.2)$ satisfies the equation of the circle.

**31.** 15.5 feet

### PART III

**32.** See the *Answers Explained* section.

**33.** $B'(7, 1)$. $\triangle DEF \cong \triangle A'B'C'$ because $\triangle DEF$ is a reflection of $\triangle A'B'C'$ over the line $x = -1$.

**34.** $34.7°$

### PART IV

**35.** See the *Answers Explained* section.

**36.** $44.53

In **PARTS II–IV** you are required to show how you arrived at your answers. For sample methods of solutions, see the *Answers Explained* section.

# Answers Explained

## PART I

1.  Two lines intersected by a transversal are parallel if the same side exterior angles or if the same side interior angles are supplementary. In the given figure, the same side exterior angles and same side interior angles formed by lines $\ell$, $p$, and $r$ are supplementary since $112° + 68° = 180°$. Therefore lines $\ell$ and $p$ are parallel.

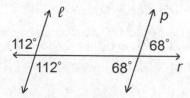

    The correct choice is (**2**).

2.  Translations, rotations, and reflections are rigid motions that always result in a congruent image. A dilation can change lengths, so the image is not necessarily congruent to the original.

    The correct choice is (**2**).

3.  A median of a triangle is a segment from a vertex to the midpoint of the opposite side. In the accompanying figure of equilateral triangle $\triangle ABC$, point $D$ is the midpoint of side $\overline{AB}$, and $\overline{CD}$ is a median. When $\triangle ABC$ is rotated about median $\overline{CD}$, a cone is generated as shown in the figure.

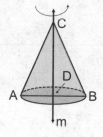

    The correct choice is (**1**).

**4.** Each of the unknown angles must be calculated before checking the possible choices. Starting with $\triangle BCD$, apply the angle sum theorem to find $\angle C$:

$$m\angle BDC + m\angle DBC + m\angle C = 180°$$
$$100° + 30° + m\angle C = 180°$$
$$130° + m\angle C = 180°$$
$$m\angle C = 50°$$

Next find $m\angle ADB$ using the linear pair relationship with $\angle BDC$:

$$m\angle ADB + m\angle BDC = 180°$$
$$m\angle ADB + 100° = 180°$$
$$m\angle ADB = 80°$$

Since $m\angle A = m\angle C = 50°$, $\triangle ABC$ has 2 congruent angles, and the opposite sides must be congruent. So $\triangle ABC$ must be isosceles.

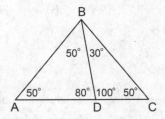

The correct choice is (**2**).

**5.** A 90° rotation of point $P$ will result in a 90° angle between $P$, the origin, and the image. Point $A$ will complete the required 90° angle.

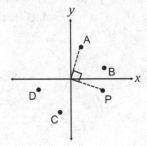

The correct choice is (**1**).

6. The two acute angles in a right triangle must be complementary, so we know that $\angle A$ and $\angle B$ are complementary. The cofunction relationship in trigonometry states that if two angles are complementary, the sine of one equals the cosine of the other. Therefore we know $\sin B = \cos A = \dfrac{\sqrt{21}}{5}$. The correct choice is **(1)**.

7. Evaluate each choice:

   - Choice (1): Opposite sides are congruent and parallel. This is sufficient to prove $ABCD$ is a parallelogram.

   - Choice (2): Both pairs of opposite sides are congruent. This is sufficient to prove $ABCD$ is a parallelogram.

   - Choice (3): One pair of sides is congruent, and the other pair of sides is parallel. This is not sufficient to prove $ABCD$ is a parallelogram since any isosceles trapezoid would have this property.

   - Choice (4): Both pairs of opposite sides are parallel. This is sufficient to prove $ABCD$ is a parallelogram.

   The correct choice is **(3)**.

8. The strategy is to sketch an altitude to one of the sides of the triangle, which also bisects a side of the triangle since two congruent triangles are formed. The new right triangle has one leg of length 10 and a hypotenuse of length 20. Apply the Pythagorean theorem to find the height $h$.

$$a^2 + b^2 = c^2$$
$$10^2 + h^2 = 20^2$$
$$100 + h^2 = 400$$
$$h^2 = 300$$
$$h = \sqrt{300}$$
$$= 17.3$$

The correct choice is **(3)**.

9. Since the transformation requires a change in size, we know a dilation is needed. So choices (1) and (2) can be eliminated. The rotation must map $\overline{EC}$ onto $\overline{EF}$. We know $\angle FEC$ is a right angle so the rotation must be 90°, eliminating choice (3).

The correct choice is (4).

10. An altitude to a hypotenuse of a right triangle is the mean proportional to the two parts of the hypotenuse:

$$\frac{AD}{CD} = \frac{CD}{BD} \qquad \text{Hypotenuse is the mean proportional}$$

$$\frac{4}{CD} = \frac{CD}{6}$$

$$CD^2 = 24 \qquad \text{Cross multiply}$$

$$CD = \sqrt{24}$$

Now we can apply the Pythagorean theorem in $\triangle ADC$ to find $AC$:

$$a^2 + b^2 = c^2$$

$$AD^2 + CD^2 = AC^2$$

$$4^2 + \left(\sqrt{24}\right)^2 = AC^2$$

$$16 + 24 = AC^2$$

$$40 = AC^2$$

$$AC = \sqrt{40}$$

Simplify the radical by factoring out the perfect square factor from 40.

$$AC = \sqrt{4 \cdot 10}$$

$$= 2\sqrt{10}$$

The correct choice is (2).

**11.** A bisector passes through the midpoint of another segment. So point $E$ is the midpoint of $\overline{AB}$, making $\overline{AE} \cong \overline{BE}$. Also, any point on perpendicular bisector $\overline{CD}$ is equidistant from the endpoints of $\overline{AB}$. Therefore, $\overline{AD} \cong \overline{BD}$ and $\overline{AC} \cong \overline{BC}$. The only segments that do not have to be congruent are $\overline{DE}$ and $\overline{CE}$ because the bisector is not necessarily bisected itself.

The correct choice is **(4)**.

**12.** A good strategy is to sketch and label the two triangles separately so the similarity relationship can be easily identified. $\angle R \cong \angle R$ by the reflexive property, and $\angle H \cong \angle RDO$ is given. The two triangles are similar by AA, and the similarity statement is $\triangle RHC \sim \triangle RDO$.

By using the similarity statement, we can write a proportion using corresponding sides to find length $CR$. $RD$ and $RO$ are given, and $HR$ is the sum of $RO$ and $OH$:

$$\frac{RH}{RD} = \frac{CR}{RO}$$

$$\frac{10}{4} = \frac{CR}{6}$$

$$4 \cdot CR = 60$$

$$CR = \frac{60}{4}$$

$$= 15$$

From the original figure, $CD = CR - RD$:

$$CD = 15 - 4$$

$$= 11$$

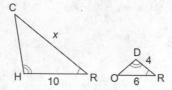

The correct choice is **(3)**.

**13.** A regular pyramid is shown in the accompanying figure with altitude $\overline{CD}$. Cross section $ABC$ contains the altitude and is a triangle.

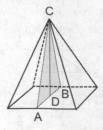

The correct choice is **(3)**.

**14.** The diagonals of a rhombus are perpendicular. So diagonal $\overline{EM}$ must be contained on the line perpendicular to $y = -x + 3$. The slope of diagonal $\overline{EM}$ is $-1$. The slope of the perpendicular line is the negative reciprocal, or $+1$. The $y$-intercept of the perpendicular line is found using $y = mx + b$, with slope $m = 1$ and the point $(2, 1)$ for $x$ and $y$.

$$y = mx + b$$
$$1 = 1(2) + b$$
$$1 = 2 + b$$
$$-1 = b$$

The perpendicular line has a slope of 1 and a $y$-intercept of $-1$. Its equation is $y = x - 1$.

The correct choice is **(1)**.

**15.** Graphing points $A$ and $B$ reveals that one leg of the triangle is vertical and has a length of 8. The required height of the triangle can be calculated from the area formula of a triangle:

$$A = \frac{1}{2} \text{ base} \cdot \text{height}$$

$$24 = \frac{1}{2} \cdot 8 \cdot \text{height}$$

$$\text{height} = 6$$

Plot each of the choices to determine which gives a triangle with a height of 6. Point $(-3, 8)$ results in a triangle with a height of 6 and an area equal to 24.

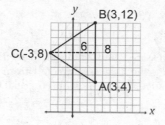

The correct choice is **(3)**.

16. Use the completing the square procedure to rewrite the equation in the form $(x - h)^2 + (y - k)^2 = r^2$. First group the $x$-terms and $y$-terms on the left. Then move the constant term to the right:

$$x^2 + y^2 - 4x + 8y + 11 = 0$$
$$x^2 - 4x + y^2 + 8y = -11$$

Next find the constant terms needed to complete the square:

$$\text{constant for the } x\text{-terms} = \left(\frac{1}{2} \cdot \text{coefficient of } x\right)^2$$

$$= \left(\frac{1}{2} \cdot (-4)\right)^2$$

$$= (-2)^2$$

$$= 4$$

$$\text{constant for the } y\text{-terms} = \left(\frac{1}{2} \cdot \text{coefficient of } y\right)^2$$

$$= \left(\frac{1}{2} \cdot 8\right)^2$$

$$= (4)^2$$

$$= 16$$

Add the required constants to each side of the equation, and factor the left side:

$$x^2 - 4x + 4 + y^2 + 8y + 16 = -11 + 4 + 16$$

$$x^2 - 4x + 4 + y^2 + 8y + 16 = 9$$

$$(x - 2)^2 + (y + 4)^2 = 9$$

The values of $h$ and $k$ are 2 and –4, respectively. Be careful with the signs because $h$ and $k$ are the values subtracted from $x$ and $y$. The center is located at (2, –4). Since $r^2 = 9$, the radius is equal to 3.

The correct choice is (**1**).

17. First find the volume by modeling the tree trunk as a cylinder with a height of 8 meters and a circumference of 4.5 meters. Use the circumference to calculate the radius:

$$\text{circumference} = 2\pi r$$

$$\frac{\text{circumference}}{2\pi} = r$$

$$\frac{4.5 \text{ m}}{2\pi} = r$$

$$r = 0.7161972 \text{ m}$$

Now calculate the volume of the cylinder:

$$V = \pi r^2 h$$

$$V = \pi (0.7161972 \text{ m})^2 (8 \text{ m})$$

$$= 12.891548 \text{ m}^3$$

Calculate the mass using the density and volume:

$$\text{mass} = \text{density} \cdot \text{volume}$$

$$\text{mass} = 752 \ \frac{\text{kg}}{\text{m}^3} \cdot 12.891548 \text{ m}^3$$

$$= 9694 \text{ kg}$$

The correct choice is (**2**).

**18.** Find the $x$- and $y$-coordinates of point $P$ using the following formula:

$$\text{ratio} = \frac{x - x_1}{x_2 - x} \;,\; \text{ratio} = \frac{y - y_1}{y_2 - y}$$

Using the coordinates $x(-6, -2)$ and $y(6, 7)$, $x_1 = -6$, $x_2 = 6$, $y_1 = -2$, and $y_2 = 7$. The ratio is $\frac{1}{5}$.

Solve for the $x$-coordinate:

$$\text{ratio} = \frac{x - x_1}{x_2 - x}$$

$$\frac{1}{5} = \frac{x - (-6)}{6 - x}$$

$$5(x - (-6)) = 1(6 - x) \qquad \text{Cross multiply}$$

$$5x + 30 = 6 - x$$

$$6x + 30 = 6$$

$$6x = -24$$

$$x = -4$$

Solve for the $y$-coordinate:

$$\text{ratio} = \frac{y - y_1}{y_2 - y}$$

$$\frac{1}{5} = \frac{y - (-2)}{7 - y}$$

$$5(y - (-2)) = 1(7 - y)$$

$$5y + 10 = 7 - y$$

$$6y + 10 = 7$$

$$6y = -3$$

$$y = -\frac{1}{2}$$

The coordinates of point $P$ are $\left(-4, -\frac{1}{2}\right)$

The correct choice is **(4)**.

**19.** The formula for the area of a sector is $A = \dfrac{\theta}{360} \cdot \pi \cdot r^2$, where $\theta$ is the central angle that intercepts the sector and $r$ is the radius of the circle. From the figure $OA$, $OB$, and $OC$ are radii, and $\theta = 108°$. Substitute these into the area formula to get

$$A = \frac{108}{360} \cdot \pi \cdot (OA)^2, \; A = \frac{108}{360} \cdot \pi \cdot (OB)^2, \; \text{or } A = \frac{108}{360} \cdot \pi \cdot (BC)^2$$

Beth is the only correct person since none of the other formulas are equivalent to these.

There was a typographical error in this Regents exam and the correct answer did not appear in the choices provided. All students received credit for this question.

**20.** The can will need a radius equal to the radius of the ball:

$$r = \frac{1}{2} \cdot \text{diameter}$$

$$= \frac{1}{2}(6.7 \text{ cm})$$

$$= 3.35 \text{ cm}$$

The height of the can must be the height of 4 balls. The height of each ball is equal to its diameter:

$$h = 4 \cdot \text{diameter}$$

$$= 4(6.7 \text{ cm})$$

$$= 26.8 \text{ cm}$$

Now apply the formula for the volume of a cylinder:

$$V = \pi r^2 h$$

$$= \pi (3.35 \text{ cm})^2 (26.8 \text{ cm})$$

$$= 945 \text{ cm}^3$$

The correct choice is **(4)**.

**21.** The length $A'B'$ can be found with the distance formula:

$$d = \sqrt{(x_1 - x_2)^2 + (y_1 - y_2)^2}$$

$$A'B' = \sqrt{(4 - 16)^2 + (-2 - 14)^2}$$

$$= \sqrt{(-12)^2 + (-16)^2}$$

$$= \sqrt{400}$$

$$= 20$$

A dilation of a segment will multiply the length by the scale factor:

preimage length · scale factor = image length

$$AB \cdot \text{scale factor} = A'B'$$

$$AB \cdot \frac{1}{2} = 20$$

$$AB = \frac{20}{\frac{1}{2}}$$

$$= 40$$

The correct choice is **(4)**.

**22.** Sketch and label the two triangles separately as shown in figure (a). We see that the SAS congruence postulate requires $AB \cong BC$. This is not one of the choices. Look at the original figure shown in figure (b). We see that if $\overline{AD} \cong \overline{CE}$, then $AD + DB = CE + EB$. This statement allows us to conclude that $\overline{AB} \cong \overline{BC}$, and the SAS postulate can be used.

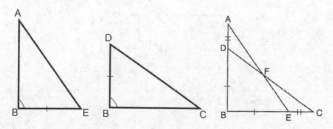

The correct choice is **(3)**.

**23.** Any point $D$ on the circle would form inscribed angle $\angle BDC$, which intercepts diameter $\overline{BAC}$. An inscribed angle that intercepts a diameter is a right angle, so $\triangle BCD$ must be a right triangle.

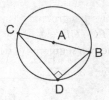

The correct choice is (**1**).

**24.** Opposite angles of a parallelogram are congruent, so m$\angle ABC$ = 112°. $\angle ABC$ and $\angle ECB$ are a linear pair, so their sum must equal 180°:

$$\text{m}\angle EBC + \text{m}\angle ABC = 180°$$
$$\text{m}\angle EBC + 112° = 180°$$
$$\text{m}\angle EBC = 68°$$

$\triangle BCE$ is isosceles with congruent legs $\overline{BE} \cong \overline{CE}$. So

$$\text{m}\angle BCE = \text{m}\angle EBC = 68°$$

Finally, apply the triangle angle sum theorem to $\triangle BCE$:

$$\text{m}\angle E + \text{m}\angle EBC + \text{m}\angle BCE = 180°$$
$$\text{m}\angle E + 68° + 68° = 180°$$
$$\text{m}\angle E + 136° = 180°$$
$$\text{m}\angle E = 44°$$

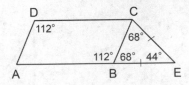

The correct choice is (**1**).

## PART II

**25.** $\overline{AC} \cong \overline{BC}$ and $\overline{CE} \cong \overline{CD}$ because two tangents to a circle from the same point (point $C$) are always congruent. Given that $AC:CE = 5:3$, we can apply substitution to get $BC : CD = 5 : 3$. By letting $\overline{BC} = 5x$ and $\overline{CD} = 3x$, we can use the sum of the two segments to write and solve an equation:

$$5x + 3x = 56$$
$$8x = 56$$
$$x = 7$$

By substituting, we find the length of $\overline{CD} = 3(7) = \textbf{21}$.

**26.** A translation 5 right and 2 up adds 5 to each $x$-coordinate and adds 2 to each $y$-coordinate. On a graph, the figure slides 5 units right and 2 units up. A reflection over the line $y = 0$ multiplies each $y$-coordinate by $-1$. On the graph, the figure flips over the line $y = 0$, which is the $x$-axis.

|  | Translate | Reflect |
|---|---|---|
| $A(1, 1)$ | $\rightarrow \quad A'(6, 3)$ | $\rightarrow \quad A''(6, -3)$ |
| $B(4, 1)$ | $\rightarrow \quad B'(9, 3)$ | $\rightarrow \quad B''(9, -3)$ |
| $C(4, 5)$ | $\rightarrow \quad C'(9, 7)$ | $\rightarrow \quad C'''(9, -7)$ |

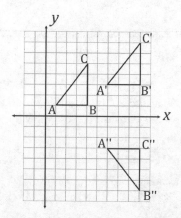

The coordinates of $\triangle A''B''C''$ are $\textbf{A}''(\textbf{6, -3})$, $\textbf{B}''(\textbf{9, -3})$, $\textbf{C}''(\textbf{9, -7})$.

**27.** The minimum rotation angle that rotates a regular polygon onto itself is $\frac{360}{n}$, where $n$ is the number of sides. A hexagon has 6 sides, so $n = 6$:

$$\text{Minimum angle} = \frac{360}{6}$$
$$= 60°$$

The minimum number of degrees in the rotatation such that the hexagon will coincide with itself is **60°**.

**28.** A median is a segment from a vertex to the opposite midpoint. So construct the midpoint of $\overline{AB}$ as follows:

- Step 1: With the point of the compass point at $A$ and the compass opened more than half the length of $AB$, make an arc. Without changing the compass setting, make a second arc but with the compass point at $B$. The arcs should cross as shown in figure (a).

- Step 2: Connect the points of intersection of the arcs to locate the midpoint $D$ as shown in figure (b).

- Step 3: Connect $C$ to $D$ to create median $\overline{CD}$ as shown in figure (c).

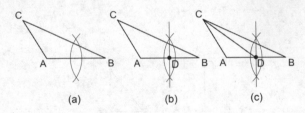

(a)                 (b)                 (c)

**29.** $m\angle P = m\angle L = 47°$ because $\angle P$ is the image of $\angle L$ and rotations are rigid motions that preserve angle measure. Next apply the angle sum theorem to $\triangle MNP$:

$$m\angle M + m\angle N + m\angle P = 180°$$
$$m\angle M + 57° + 47° = 180°$$
$$m\angle M + 104° = 180°$$
$$m\angle M = 76°$$

The measure of angle $M$ is **76°**.

**30.** The equation of a circle is $(x - h)^2 + (y - k)^2 = r^2$, where $(h, k)$ are the coordinates of the center and $r$ is the radius. Substitute $(1, -2)$ for $h$ and $k$, 4 for $r$, and the coordinates $(3.4, 1.2)$ for $x$ and $y$. If the equation is balanced, then $(3.4, 1.2)$ does lie on the circle:

$$(x - h)^2 + (y - k)^2 = r^2$$

$$(3.4 - 1)^2 + (1.2 - (-2))^2 \overset{?}{=} 4^2$$

$$(2.4)^2 + (3.2)^2 \overset{?}{=} 4^2$$

$$5.76 + 10.24 \overset{?}{=} 16$$

$$16 = 16$$

Since the equation is balanced, **the point (3.4, 1.2) does lie on the circle**.

**31.** The wall, ladder, and ground form a right triangle. A trigonometric ratio can be used to find the length of the ladder. Relative to the 75° angle, the 15-foot wall is the opposite side and the ladder is the hypotenuse. We can use the sine ratio to find the length of the ladder, $x$:

$$\sin 75° = \frac{\text{opposite}}{\text{hypotenuse}}$$

$$\sin 75° = \frac{15}{x}$$

$$x \sin 75° = 15$$

$$\sin 75° = \frac{15}{\sin 75°}$$

$$= 15.52914271$$

To the *nearest tenth of a foot*, the ladder is **15.5** feet long.

# PART III

**32.** Place the compass point at $B$, and make an arc that passes through $A$. With the same compass opening, move the compass point to $A$ and make another arc. Repeat these steps using points $B$ and $C$, as shown in figure (a). Extend $\overline{BA}$ to $A'$ and extend $\overline{BC}$ to $C'$, as shown in figure (b). $\triangle A'B'C'$ is a dilation by a scale factor of 2 centered at $B$.

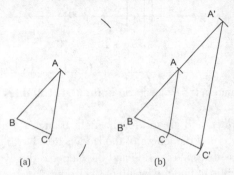

**$\overline{A'C'}$ is twice as long as $\overline{AC}$** because all pairs of corresponding sides are in the same ratio.

**33.** Points $C$, $A$, and $C'$ form a right angle. So the rotation is 90°. Therefore, every side of $\triangle A'B'C'$ will have a slope equal to the negative reciprocal of its corresponding side. The lengths of corresponding sides will be equal because rotations are rigid motions:

$$\text{Slope of } \overline{BC} = \frac{\text{rise}}{\text{run}}$$

$$= \frac{1}{4}$$

$$\text{Slope of } \overline{B'C'} = -\frac{4}{1}$$

To find the coordinates of $B'$, apply a rise of 4 and a run of –1 from $C'$. By counting up 4 and left 1 from $C'$ (8, –3), we find **$B'$ has coordinates (7, 1)**. Note that if you had switched the sign and used a rise of –4 and a run of 1, point $B'$ would have ended up on the wrong side of $C'$.

$\triangle DEF$ is a reflection of $\triangle A'B'C'$ over the line $x = -1$. $E$ and $B'$ are both 8 units from $x = -1$. $D$ and $A'$ are both 3 units from $x = -1$. $F$ and $BC$ are both 9 units from $x = -1$. $\triangle DEF \cong \triangle A'B'C'$ because a reflection is a rigid motion.

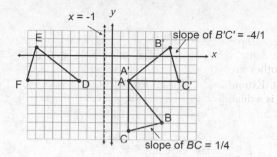

The location of $B'$ is $(7, 1)$. **$\triangle DEF$ is congruent to $\triangle A'B'C'$.**

**34.** The projection angle can be found using the two right triangles $\triangle ABC$ and $\triangle ABD$ and trigonometric ratios.

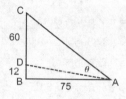

First calculate $m\angle BAD$ using the opposite side $\overline{BD}$, the adjacent side $\overline{AB}$, and the tangent ratio:

$$\tan BAD = \frac{\text{opposite}}{\text{adjacent}}$$

$$\tan BAD = \frac{BD}{AB}$$

$$\tan BAD = \frac{12}{75}$$

$$m\angle BAD = \tan^{-1}\frac{12}{75} \quad \text{Use the inverse tangent to find the angle}$$

$$= 9.090276°$$

Next find m∠BAC using △ABC:

$$\tan BAC = \frac{\text{opposite}}{\text{adjacent}}$$

$$\tan BAC = \frac{BC}{AB}$$

$$\tan BAC = \frac{BD + DC}{AB}$$

$$\tan BAC = -\frac{12 + 60}{75}$$

$$m\angle BAC = \tan^{-1}\frac{72}{75}$$

$$= 43.830860°$$

From the two triangles, we can use angle subtraction to find $\theta$:

$$\theta = m\angle BAC - m\angle BAD$$

$$= 43.830860° - 9.090276°$$

$$= 34.740584$$

*To the nearest tenth of a degree, $\theta$ = **34.7°**.*

**35.** The strategy is to construct two auxiliary segments, $\overline{BC}$ and $\overline{AD}$, and then prove that $\triangle ADE$ and $\triangle CBE$ are similar. From there, we can apply the theorem that corresponding parts are proportional to write the ratio $\dfrac{AE}{CE} = \dfrac{ED}{EB}$.

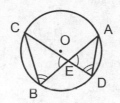

| Statements | Reasons |
|---|---|
| 1. Circle $O$, chords $\overline{AB}$ and $\overline{CD}$ intersect at $E$ | 1. Given |
| 2. Construct $\overline{BC}$ and $\overline{AD}$ | 2. A segment can be constructed between two endpoints |
| 3. $\angle BEC \cong \angle DEA$ | 3. Vertical angles are congruent |
| 4. $\angle B \cong \angle D$ | 4. Inscribed angles that intercept the same arc are congruent |
| 5. $\triangle ADE \sim CBE$ | 5. AA |
| 6. $\dfrac{AE}{CE} = \dfrac{ED}{EB}$ | 6. Corresponding parts of similar triangles are proportional |
| 7. $AE \cdot EB = CE \cdot ED$ | 7. Cross products are equal in a proportion |

**36.** To solve this problem, follow these steps:

- Step 1: Find the volume of the cone.

- Step 2: Find the volume of the hemisphere.

- Step 3: Find the total volume.

- Step 4: Find the mass of 1 cone.

- Step 5: Find the cost of ice for 1 cone.

- Step 6: Find the cost of ice for 50 cones.

Find the volume of the cone:

$$\text{Radius} = \frac{1}{2} \cdot \text{diameter}$$

$$= \frac{1}{2}(8.3)$$

$$= 4.15$$

$$V_{\text{cone}} = \frac{1}{3}\pi r^2 h$$

$$= \frac{1}{3}\pi(4.15\text{ cm})^2(10.2\text{ cm})$$

$$= 183.960670\text{ cm}^3$$

Find the volume of the hemisphere. Use the formula for a sphere and multiply it by $\frac{1}{2}$:

$$V_{\text{hemisphere}} = \frac{4}{3}\pi r^3 \cdot \frac{1}{2}$$

$$= \frac{4}{3}\pi(4.15\text{ cm})^3 \cdot \frac{1}{2}$$

$$= 149.693486\text{ cm}^3$$

Find the total volume:

$$V = V_{\text{cone}} + V_{\text{hemisphere}}$$

$$= 183.960670\text{ cm}^3 + 149.693486\text{ cm}^3$$

$$= 333.654156\text{ cm}^3$$

Find the mass of 1 cone:

$$\text{Mass} = \text{volume} \cdot \text{density}$$

$$= 333.654156 \text{ cm}^3 \cdot 0.697 \frac{\text{grams}}{\text{cm}^3}$$

$$= 232.556947 \text{ grams}$$

Find the cost of ice for 1 cone. First convert the cost from dollars per kilogram to dollars per gram. Remember that there are 1000 grams per kilogram:

$$\frac{\$3.83}{\text{kg}} \cdot \frac{1 \text{ kg}}{1000 \text{ grams}} = \frac{\$0.00383}{\text{gram}}$$

$$\text{cost}_{1 \text{ cone}} = \text{mass} \cdot \frac{\text{cost}}{\text{gram}}$$

$$= 232.556947 \text{ grams} \cdot \frac{\$0.00383}{\text{gram}}$$

$$= \$0.890693$$

Find the cost of ice for 50 cones:

$$\text{cost}_{50 \text{ cones}} = \text{cost}_{1 \text{ cone}} \cdot 50$$

$$= \$0.890693 \cdot 50$$

$$= \$44.534655$$

To the nearest cent, the cost of ice for making 50 snow cones is **$44.53**.

| Topic | Question Numbers | Number of Points | Your Points | Your Percentage |
|---|---|---|---|---|
| 1. Basic Angle and Segment Relationships | 1, 11 | 2 + 2 = 4 | | |
| 2. Angle and Segment Relationships in Triangles and Polygons | 4, 8 | 2 + 2 = 4 | | |
| 3. Constructions | 28, 32 | 2 + 4 = 6 | | |
| 4. Transformations | 2, 5, 9, 26, 27, 29, 33a | 2 + 2 + 2 + 2 + 2 + 2 + 2 = 14 | | |
| 5. Triangle Congruence | | | | |
| 6. Lines, Segments, and Circles on the Coordinate Plane | 14, 15, 16, 18, 21, 30 | 2 + 2 + 2 + 2 + 2 + 2 = 12 | | |
| 7. Similarity | 10, 12, 22 | 2 + 2 + 2 = 6 | | |
| 8. Trigonometry | 6, 31, 34 | 2 + 2 + 4 = 8 | | |
| 9. Parallelograms | 7, 24 | 2 + 2 = 4 | | |
| 10. Coordinate Geometry Proofs | 33b | 2 | | |
| 11. Volume and Solids | 3, 13, 20 | 2 + 2 + 2 = 6 | | |
| 12. Modeling | 17, 36 | 2 + 6 = 8 | | |
| 13. Circles | 19, 23, 25, 35 | 2 + 2 + 2 + 6 = 12 | | |

# HOW TO CONVERT YOUR RAW SCORE TO YOUR GEOMETRY REGENTS EXAMINATION SCORE

The conversion chart below can be used to determine your final score on the August 2016 Regents Exam in Geometry. To find your final exam score, locate in the "Raw Score" column the total number of points you scored out of a possible 86. Then locate in the adjacent column to the right the scale score that corresponds to your raw score. The scale score is your final score.

| Raw Score | Scale Score | Performance Level | Raw Score | Scale Score | Performance Level | Raw Score | Scale Score | Performance Level |
|---|---|---|---|---|---|---|---|---|
| 86 | 100 | 5 | 57 | 79 | 3 | 28 | 59 | 2 |
| 85 | 99 | 5 | 56 | 78 | 3 | 27 | 58 | 2 |
| 84 | 97 | 5 | 55 | 78 | 3 | 26 | 57 | 2 |
| 83 | 96 | 5 | 54 | 77 | 3 | 25 | 55 | 2 |
| 82 | 95 | 5 | 53 | 77 | 3 | 24 | 54 | 1 |
| 81 | 94 | 5 | 52 | 77 | 3 | 23 | 53 | 1 |
| 80 | 93 | 5 | 51 | 76 | 3 | 22 | 51 | 1 |
| 79 | 92 | 5 | 50 | 76 | 3 | 21 | 49 | 1 |
| 78 | 91 | 5 | 49 | 75 | 3 | 20 | 48 | 1 |
| 77 | 91 | 5 | 48 | 75 | 3 | 19 | 46 | 1 |
| 76 | 90 | 5 | 47 | 74 | 3 | 18 | 44 | 1 |
| 75 | 89 | 5 | 46 | 74 | 3 | 17 | 43 | 1 |
| 74 | 88 | 5 | 45 | 73 | 3 | 16 | 41 | 1 |
| 73 | 87 | 5 | 44 | 72 | 3 | 15 | 39 | 1 |
| 72 | 87 | 5 | 43 | 72 | 3 | 14 | 37 | 1 |
| 71 | 86 | 5 | 42 | 71 | 3 | 13 | 35 | 1 |
| 70 | 86 | 5 | 41 | 71 | 3 | 12 | 32 | 1 |
| 69 | 85 | 5 | 40 | 70 | 3 | 11 | 30 | 1 |
| 68 | 84 | 4 | 39 | 69 | 3 | 10 | 28 | 1 |
| 67 | 84 | 4 | 38 | 69 | 3 | 9 | 25 | 1 |
| 66 | 83 | 4 | 37 | 68 | 3 | 8 | 23 | 1 |
| 65 | 83 | 4 | 36 | 67 | 3 | 7 | 20 | 1 |
| 64 | 82 | 4 | 35 | 66 | 3 | 6 | 18 | 1 |
| 63 | 82 | 4 | 34 | 65 | 3 | 5 | 15 | 1 |
| 62 | 81 | 4 | 33 | 64 | 2 | 4 | 12 | 1 |
| 61 | 81 | 4 | 32 | 63 | 2 | 3 | 9 | 1 |
| 60 | 80 | 4 | 31 | 62 | 2 | 2 | 6 | 1 |
| 59 | 80 | 4 | 30 | 61 | 2 | 1 | 3 | 1 |
| 58 | 79 | 3 | 29 | 60 | 2 | 0 | 0 | 1 |

# Examination June 2017
## Geometry (Common Core)

## GEOMETRY REFERENCE SHEET

| | |
|---|---|
| 1 inch = 2.54 centimeters | 1 ton = 2000 pounds |
| 1 meter = 39.37 inches | 1 cup = 8 fluid ounces |
| 1 mile = 5280 feet | 1 pint = 2 cups |
| 1 mile = 1760 yards | 1 quart = 2 pints |
| 1 mile = 1.609 kilometers | 1 gallon = 4 quarts |
| 1 kilometer = 0.62 mile | 1 gallon = 3.785 liters |
| 1 pound = 16 ounces | 1 liter = 0.264 gallon |
| 1 pound = 0.454 kilogram | 1 liter = 1000 cubic centimeters |
| 1 kilogram = 2.2 pounds | |

| | |
|---|---|
| Triangle | $A = \frac{1}{2}bh$ |
| Parallelogram | $A = bh$ |
| Circle | $A = \pi r^2$ |
| Circle | $C = \pi d$ or $C = 2\pi r$ |
| General Prisms | $V = Bh$ |
| Cylinder | $V = \pi r^2 h$ |
| Sphere | $V = \frac{4}{3}\pi r^3$ |

| Cone | $V = \dfrac{1}{3}\pi r^2 h$ |
|---|---|
| Pyramid | $V = \dfrac{1}{3}Bh$ |
| Pythagorean Theorem | $a^2 + b^2 = c^2$ |
| Quadratic Formula | $x = \dfrac{-b \pm \sqrt{b^2 - 4ac}}{2a}$ |
| Arithmetic Sequence | $a_n = a_1 + (n-1)d$ |
| Geometric Sequence | $a_n = a_1 r^{n-1}$ |
| Geometric Series | $S_n = \dfrac{a_1 - a_1 r^n}{1-r}$ where $r \neq 1$ |
| Radians | 1 radian $= \dfrac{180}{\pi}$ degrees |
| Degrees | 1 degree $= \dfrac{\pi}{180}$ radians |
| Exponential Growth/Decay | $A = A_0 e^{k(t - t_0)} + B_0$ |

## PART I

Answer all 24 questions in this part. Each correct answer will receive 2 credits. No partial credit will be allowed. For each statement or question, write in the space provided the numeral preceding the word or expression that best completes the statement or answers the question. [48 credits]

**1** In the diagram below, $\triangle ABC \cong \triangle DEF$.

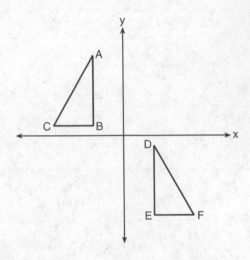

Which sequence of transformations maps $\triangle ABC$ onto $\triangle DEF$?

(1) a reflection over the x-axis followed by a translation

(2) a reflection over the y-axis followed by a translation

(3) a rotation of 180° about the origin followed by a translation

(4) a counterclockwise rotation of 90° about the origin followed by a translation

1 _____

**2** On the set of axes below, the vertices of △*PQR* have coordinates *P*(–6, 7), *Q*(2, 1), and *R*(–1, –3).

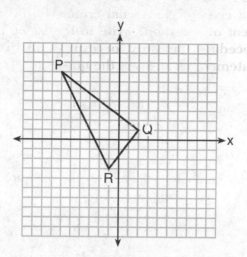

What is the area of △*PQR*?

(1) 10          (3) 25

(2) 20          (4) 50          2 _____

**3** In right triangle *ABC*, m∠*C* = 90°. If cos *B* = $\frac{5}{13}$, which function also equals $\frac{5}{13}$?

(1) tan *A*          (3) sin *A*

(2) tan *B*          (4) sin *B*          3 _____

**4** In the diagram below, m$\overset{\frown}{ABC}$ = 268°.

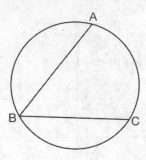

What is the number of degrees in the measure of ∠ABC?

(1) 134°                    (3) 68°

(2) 92°                     (4) 46°                    4 _____

**5** Given △MRO shown below, with trapezoid PTRO, MR = 9, MP = 2, and PO = 4.

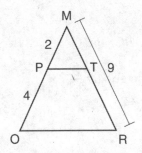

What is the length of $\overline{TR}$?

(1) 4.5                     (3) 3

(2) 5                       (4) 6                      5 _____

**6** A line segment is dilated by a scale factor of 2 centered at a point not on the line segment. Which statement regarding the relationship between the given line segment and its image is true?

(1) The line segments are perpendicular, and the image is one-half of the length of the given line segment.

(2) The line segments are perpendicular, and the image is twice the length of the given line segment.

(3) The line segments are parallel, and the image is twice the length of the given line segment.

(4) The line segments are parallel, and the image is one-half of the length of the given line segment.    6 _____

**7** Which figure always has exactly four lines of reflection that map the figure onto itself?

(1) square             (3) regular octagon

(2) rectangle         (4) equilateral triangle    7 _____

**8** In the diagram below of circle $O$, chord $\overline{DF}$ bisects chord $\overline{BC}$ at $E$.

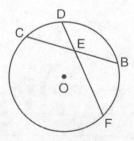

If $BC = 12$ and $FE$ is 5 more than $DE$, then $FE$ is

(1) 13            (3) 6

(2) 9             (4) 4      8 _____

**9** Kelly is completing a proof based on the figure below.

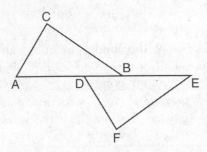

She was given that $\angle A \cong \angle EDF$, and has already proven $\overline{AB} \cong \overline{DE}$. Which pair of corresponding parts and triangle congruency method would *not* prove $\triangle ABC \cong \triangle DEF$?

(1) $\overline{AC} \cong \overline{DF}$ and SAS
(2) $\overline{BC} \cong \overline{EF}$ and SAS
(3) $\angle C \cong \angle F$ and AAS
(4) $\angle CBA \cong \angle FED$ and ASA

9 _____

**10** In the diagram below, $\overline{DE}$ divides $\overline{AB}$ and $\overline{AC}$ proportionally, m$\angle C = 26°$, m$\angle A = 82°$, and $\overline{DF}$ bisects $\angle BDE$.

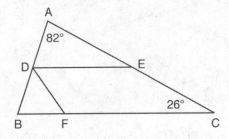

The measure of angle *DFB* is

(1) 36°      (3) 72°
(2) 54°      (4) 82°

10 _____

**11** Which set of statements would describe a parallelo-gram that can always be classified as a rhombus?

   I.  Diagonals are perpendicular bisectors of each other.

   II.  Diagonals bisect the angles from which they are drawn.

   III. Diagonals form four congruent isosceles right triangles.

(1) I and II             (3) II and III

(2) I and III           (4) I, II, and III      11 _____

**12** The equation of a circle is $x^2 + y^2 - 12y + 20 = 0$. What are the coordinates of the center and the length of the radius of the circle?

(1) center $(0, 6)$ and radius 4

(2) center $(0, -6)$ and radius 4

(3) center $(0, 6)$ and radius 16

(4) center $(0, -6)$ and radius 16      12 _____

**13** In the diagram of $\triangle RST$ below, $m\angle T = 90°$, $RS = 65$, and $ST = 60$.

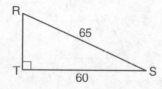

What is the measure of $\angle S$, to the *nearest degree*?

(1) 23°             (3) 47°

(2) 43°            (4) 67°         13 _____

**14** Triangle $A'B'C'$ is the image of $\triangle ABC$ after a dilation followed by a translation.

Which statement(s) would always be true with respect to this sequence of transformations?

I. $\triangle ABC \cong \triangle A'B'C'$
II. $\triangle ABC \sim \triangle A'B'C'$
III. $\overline{AB} \parallel \overline{A'B'}$
IV. $AA' = BB'$

(1) II, only
(2) I and II
(3) II and III
(4) II, III, and IV

14 _____

**15** Line segment $RW$ has endpoints $R(-4, 5)$ and $W(6, 20)$. Point $P$ is on $\overline{RW}$ such that $RP : PW$ is $2 : 3$. What are the coordinates of point $P$?

(1) $(2, 9)$
(2) $(0, 11)$
(3) $(2, 14)$
(4) $(10, 2)$

15 _____

**16** The pyramid shown below has a square base, a height of 7, and a volume of 84.

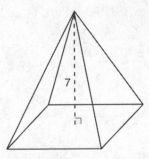

What is the length of the side of the base?

(1) 6
(2) 12
(3) 18
(4) 36

16 _____

**17** In the diagram below of triangle *MNO*, ∠*M* and ∠*O* are bisected by $\overline{MS}$ and $\overline{OR}$, respectively. Segments *MS* and *OR* intersect at *T*, and m∠*N* = 40°.

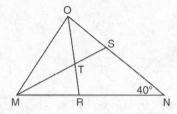

If m∠*TMR* = 28°, the measure of angle *OTS* is

(1) 40°          (3) 60°

(2) 50°          (4) 70°                    17 _____

**18** In the diagram below, right triangle *ABC* has legs whose lengths are 4 and 6.

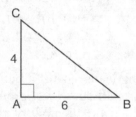

What is the volume of the three-dimensional object formed by continuously rotating the right triangle around $\overline{AB}$?

(1) 32π          (3) 96π

(2) 48π          (4) 144π                   18 _____

**19** What is an equation of a line that is perpendicular to the line whose equation is $2y = 3x - 10$ and passes through $(-6, 1)$?

(1) $y = -\dfrac{2}{3}x - 5$  (3) $y = \dfrac{2}{3}x + 1$

(2) $y = -\dfrac{2}{3}x - 3$  (4) $y = \dfrac{2}{3}x + 10$   19 _____

**20** In quadrilateral *BLUE* shown below, $\overline{BE} \cong \overline{UL}$.

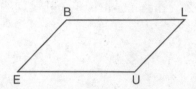

Which information would be sufficient to prove quadrilateral *BLUE* is a parallelogram?

(1) $\overline{BL} \parallel \overline{EU}$  (3) $\overline{BE} \cong \overline{BL}$
(2) $\overline{LU} \parallel \overline{BE}$  (4) $\overline{LU} \cong \overline{EU}$   20 _____

**21** A ladder 20 feet long leans against a building, forming an angle of 71° with the level ground. To the *nearest foot*, how high up the wall of the building does the ladder touch the building?

(1) 15  (3) 18
(2) 16  (4) 19   21 _____

**22** In the two distinct acute triangles $ABC$ and $DEF$, $\angle B \cong \angle E$. Triangles $ABC$ and $DEF$ are congruent when there is a sequence of rigid motions that maps

(1) $\angle A$ onto $\angle D$, and $\angle C$ onto $\angle F$
(2) $\overline{AC}$ onto $\overline{DF}$, and $\overline{BC}$ onto $\overline{EF}$
(3) $\angle C$ onto $\angle F$, and $\overline{BC}$ onto $\overline{EF}$
(4) point $A$ onto point $D$, and $\overline{AB}$ onto $\overline{DE}$          22 _____

**23** A fabricator is hired to make a 27-foot-long solid metal railing for the stairs at the local library. The railing is modeled by the diagram below. The railing is 2.5 inches high and 2.5 inches wide and is comprised of a rectangular prism and a half-cylinder.

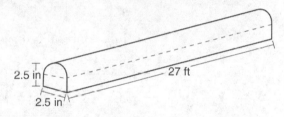

2.5 in
2.5 in
27 ft

How much metal, to the *nearest cubic inch*, will the railing contain?

(1) 151          (3) 1808
(2) 795          (4) 2025          23 _____

**24** In the diagram below, $AC = 7.2$ and $CE = 2.4$.

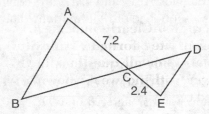

Which statement is *not* sufficient to prove $\triangle ABC \sim \triangle EDC$?

(1) $\overline{AB} \parallel \overline{ED}$
(2) $DE = 2.7$ and $AB = 8.1$
(3) $CD = 3.6$ and $BC = 10.8$
(4) $DE = 3.0$, $AB = 9.0$, $CD = 2.9$, and $BC = 8.7$     24 _____

## PART II

Answer all 7 questions in this part. Each correct answer will receive 2 credits. Clearly indicate the necessary steps, including appropriate formula substitutions, diagrams, graphs, charts, etc. For all questions in this part, a correct numerical answer with no work shown will receive only 1 credit.   [14 credits]

**25** Given: Trapezoid $JKLM$ with $\overline{JK}$ // $\overline{ML}$

Using a compass and straightedge, construct the altitude from vertex $J$ to $\overline{ML}$.
[Leave all construction marks.]

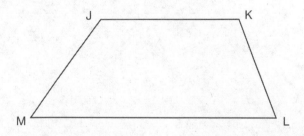

**26** Determine and state, in terms of $\pi$, the area of a sector that intercepts a 40° arc of a circle with a radius of 4.5.

**27** The diagram below shows two figures. Figure A is a right triangular prism and figure B is an oblique triangular prism. The base of figure A has a height of 5 and a length of 8 and the height of prism A is 14. The base of figure B has a height of 8 and a length of 5 and the height of prism B is 14.

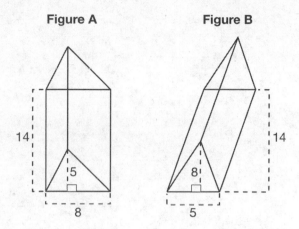

Use Cavalieri's principle to explain why the volumes of these two triangular prisms are equal.

**28** When volleyballs are purchased, they are not fully inflated. A partially inflated volleyball can be modeled by a sphere whose volume is approximately 180 in³. After being fully inflated, its volume is approximately 294 in³. To the *nearest tenth of an inch*, how much does the radius increase when the volleyball is fully inflated?

**29** In right triangle $ABC$ shown below, altitude $\overline{CD}$ is drawn to hypotenuse $\overline{AB}$. Explain why $\triangle ABC \sim \triangle ACD$.

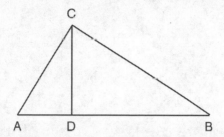

**30** Triangle *ABC* and triangle *DEF* are drawn below.

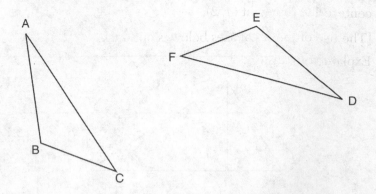

If $\overline{AB} \cong \overline{DE}$, $\overline{AC} \cong \overline{DF}$, and $\angle A \cong \angle D$, write a sequence of transformations that maps triangle *ABC* onto triangle *DEF*.

**31** Line $n$ is represented by the equation $3x + 4y = 20$. Determine and state the equation of line $p$, the image of line $n$, after a dilation of scale factor $\frac{1}{3}$ centered at the point $(4, 2)$.

[The use of the set of axes below is optional.]

Explain your answer.

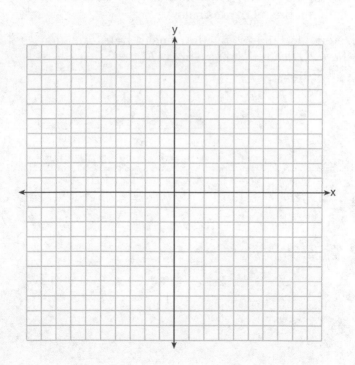

## PART III

Answer all 3 questions in this part. Each correct answer will receive 4 credits. Clearly indicate the necessary steps, including appropriate formula substitutions, diagrams, graphs, charts, etc. For all questions in this part, a correct numerical answer with no work shown will receive only 1 credit. [12 credits]

**32** Triangle $ABC$ has vertices at $A(-5, 2)$, $B(-4, 7)$, and $C(-2, 7)$, and triangle $DEF$ has vertices at $D(3, 2)$, $E(2, 7)$, and $F(0, 7)$. Graph and label $\triangle ABC$ and $\triangle DEF$ on the set of axes below.

Determine and state the single transformation where $\triangle DEF$ is the image of $\triangle ABC$.

**Question 32 is continued on the next page.**

## Question 32 continued

Use your transformation to explain why
$\triangle ABC \cong \triangle DEF$.

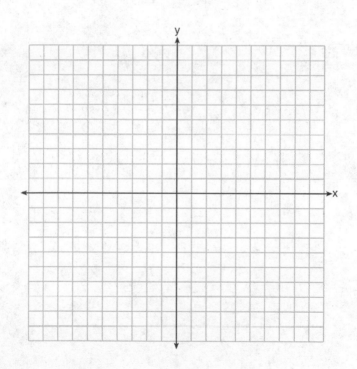

**33** Given: $\overline{RS}$ and $\overline{TV}$ bisect each other at point $X$
$\overline{TR}$ and $\overline{SV}$ are drawn

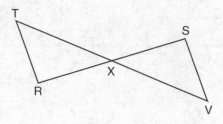

Prove: $\overline{TR} \mathbin{/\mkern-5mu/} \overline{SV}$

**34** A gas station has a cylindrical fueling tank that holds the gasoline for its pumps, as modeled below. The tank holds a maximum of 20,000 gallons of gasoline and has a length of 34.5 feet.

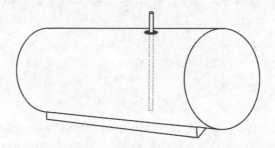

A metal pole is used to measure how much gas is in the tank. To the *nearest tenth of a foot*, how long does the pole need to be in order to reach the bottom of the tank and still extend one foot outside the tank? Justify your answer. [1 ft$^3$ = 7.48 gallons]

## PART IV

Answer the 2 questions in this part. Each correct answer will receive 6 credits. Clearly indicate the necessary steps, including appropriate formula substitutions, diagrams, graphs, charts, etc. For all questions in this part, a correct numerical answer with no work shown will receive only 1 credit. [12 credits]

**35** Quadrilateral *PQRS* has vertices *P*(–2, 3), *Q*(3, 8), *R*(4, 1), and *S*(–1, –4).

Prove that *PQRS* is a rhombus.
[The use of the set of axes on the next page is optional.]

**Question 35 is continued on the next page.**

**Question 35 continued**

Prove that *PQRS* is *not* a square.
[The use of the set of axes below is optional.]

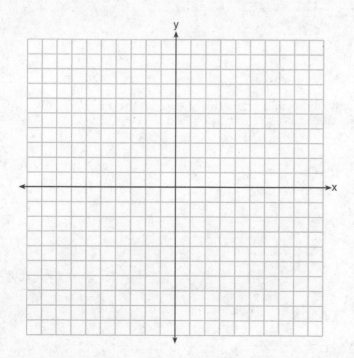

**36** Freda, who is training to use a radar system, detects an airplane flying at a constant speed and heading in a straight line to pass directly over her location. She sees the airplane at an angle of elevation of 15° and notes that it is maintaining a constant altitude of 6250 feet. One minute later, she sees the airplane at an angle of elevation of 52°. How far has the airplane traveled, to the *nearest foot*?

Determine and state the speed of the airplane, to the *nearest mile per hour*.

# Answers
# June 2017
## Geometry (Common Core)

## Answer Key

### PART I

| | | | | | |
|---|---|---|---|---|---|
| **1.** (2) | **5.** (4) | **9.** (2) | **13.** (1) | **17.** (4) | **21.** (4) |
| **2.** (3) | **6.** (3) | **10.** (2) | **14.** (*) | **18.** (1) | **22.** (**) |
| **3.** (3) | **7.** (1) | **11.** (4) | **15.** (2) | **19.** (2) | **23.** (3) |
| **4.** (4) | **8.** (2) | **12.** (1) | **16.** (1) | **20.** (2) | **24.** (2) |

*Question 14—When scoring this question, either choice 1 or choice 3 should be awarded credit.

**Question 22—When scoring this question, all students should be awarded credit regardless of the answer, if any, they record on the answer sheet for this question.

### PART II

**25.** See the complete solution for the construction.

**26.** $2.25\pi$ or $\frac{9}{4}\pi$

**27.** The two solids have the same cross sectional area and height; therefore, Cavalieri's principle states the solids must have the same volume.

**28.** 0.6 inches

**29.** See the complete solution for the proof.

**30.** Translate $\triangle ABC$ by vector $\overline{AD}$ so point $A$ maps to point $D$, then rotate $\triangle ABC$ about point $D$ so point $C$ maps to point $E$.

**31.** The image is the original line, $3x + 4y = 20$, because the center of dilation lies on the line.

### PART III

**32.** Reflection over $x = -1$. The triangles are congruent because reflections are rigid motions which preserve size and shape.

**33.** See the complete solution for the proof.

**34.** 10.9 feet

### PART IV

**35.** See the complete solution for the proof.

**36.** distance = 18442 feet,

speed = $210 \dfrac{\text{miles}}{\text{hour}}$

In **PARTS II–IV**, you are required to show how you arrived at your answers. For sample methods of solutions, see the *Answers Explained* section.

# Answers Explained

## PART I

1. We know one reflection is needed because the orientation changes between the two triangles. In $\triangle ABC$, the vertices are ordered clockwise. In $\triangle DEF$, the vertices are ordered counterclockwise. A reflection over the $y$-axis maps $\triangle ABC$ to $\triangle A'B'C'$. Next, a translation downward maps $\triangle A'B'C'$ to $\triangle DEF$.

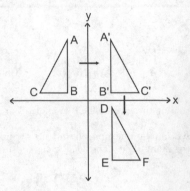

The correct choice is (**2**).

2. The area of $\triangle PQR$ can be calculated using the area of the bounding rectangle and subtracting the areas of triangles I, II, and III.

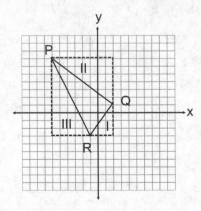

- Rectangle:

$$\text{Area} = l \cdot w$$
$$= 10 \cdot 8$$
$$= 80$$

- Triangle I:

$$\text{Area} = \frac{1}{2}b \cdot h$$
$$= \frac{1}{2}(3 \cdot 4)$$
$$= 6$$

- Triangle II:

$$\text{Area} = \frac{1}{2}b \cdot h$$
$$= \frac{1}{2}(8 \cdot 6)$$
$$= 24$$

- Triangle III:

$$\text{Area} = \frac{1}{2}b \cdot h$$
$$- \frac{1}{2}(5 \cdot 10)$$
$$= 25$$

$$\text{area } \triangle PQR = \text{area}_{\text{rectangle}} - \text{area}_{\text{triangle I}} - \text{area}_{\text{triangle II}} - \text{area}_{\text{triangle III}}$$
$$= 80 - 6 - 24 - 25$$
$$= 25$$

The correct choice is **(3)**.

3. The cofunction relationship for trigonometric ratios states that if the sum of angles $A$ and $B$ equals $90°$, then $\sin A = \cos B$. We know the sum of angles $A$ and $B$ must equal $90°$ since the third angle of the triangle measures $90°$. Therefore $\sin A = \cos B = \dfrac{5}{13}$.

The correct choice is **(3)**.

**4.** First find the measure of $\overset{\frown}{AC}$ by using the fact that the sum of the arc measures around a circle equals $360°$.

$$m\overset{\frown}{AC} + m\overset{\frown}{ABC} = 360°$$

$$m\overset{\frown}{AC} + 268° = 360°$$

$$m\overset{\frown}{AC} = 92°$$

$\angle ABC$ is an inscribed angle, which means it measures half the intercepted arc $\overset{\frown}{AC}$.

$$m\angle ABC = \frac{1}{2}m\overset{\frown}{AC}$$

$$= \frac{1}{2}(92°)$$

$$= 46°$$

The correct choice is (**4**).

**5.** Trapezoid *PTRO* has parallel sides $\overline{PT}$ and $\overline{OR}$. When a segment intersects two sides of a triangle and is parallel to the third side, it divides the triangle into similar triangles. Therefore $\triangle PTM \sim \triangle ORM$. We can set up a proportion to find the length of *MT*. Using the similarity statement:

$$\frac{MP}{MO} = \frac{MT}{MR}$$

$$\frac{2}{6} = \frac{MT}{9}$$

$$6MT = 18 \quad \text{Cross multiply}$$

$$MT = 3$$

Now calculate *TR*:

$$TR + MT = 9$$

$$TR + 3 = 9$$

$$TR = 6$$

The correct choice is (**4**).

**6.** Any dilation of a line segment will result in a parallel segment. A scale factor of 2 will make the image twice as long.

The correct choice is (**3**).

**7.** A line of reflection maps a figure onto itself if you can fold the figure over the line and the two halves sit exactly on top of each other. The only figure among the choices that has 4 lines of reflection is the square.

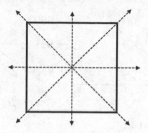

The correct choice is **(1)**.

**8.** When two chords intersect, the products of the parts are equal. In this case, $DE \cdot FE = BE \cdot CE$. Since $BC$ is 12 and it is bisected, we know that $BE$ and $CE$ are each 6. Let $DE = x$ and $FE = x + 5$:

$$DE \cdot FE = BE \cdot CE$$
$$x(x + 5) = 6 \cdot 6$$
$$x^2 + 5x = 36$$
$$x^2 + 5x - 36 = 0$$

Factor and set each factor equal to zero:

$$(x + 9)(x - 4) = 0$$
$$(x + 9) = 0 \quad \text{and} \quad (x - 4) = 0$$
$$x = -9 \qquad\qquad x = 4$$

Since lengths cannot be negative, we know $x = 4$ and $DE = 4 + 5 = 9$.

The correct choice is **(2)**.

9. A good method for dealing with overlapping triangles is to sketch the triangles separately and mark each figure with congruent markings. The figure below shows the triangles and markings for choice (2). Two pairs of sides and one pair of angles are congruent. However, the angle is not the included angle. So SAS does not apply.

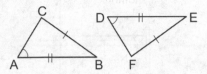

The correct choice is **(2)**.

10. A segment that divides two sides of a triangle proportionally is parallel to the third side. Therefore, $\overline{DE} \parallel \overline{BFC}$.

∠C and ∠DEA are congruent corresponding angles. So

$$m\angle DEA = m\angle C = 26°$$

Use the angle sum theorem in △ADE to find ∠ADE:

$$m\angle ADE + m\angle DEA + m\angle A = 180°$$
$$m\angle ADE + 26° + 82° = 180°$$
$$m\angle ADE + 108° = 180°$$
$$m\angle ADE = 72°$$

∠BDE and ∠ADE are a linear pair that sum to 180°:

$$m\angle BDE + m\angle ADE = 180°$$
$$m\angle BDE = 180° - m\angle ADE$$
$$= 180° - 72°$$
$$= 108°$$

Since $\overline{DF}$ bisects ∠BDE, we can find m∠FDE:

$$m\angle FDE = \frac{1}{2}m\angle BDE$$

$$= \frac{1}{2}(108°)$$

$$= 54°$$

∠DFB and ∠FDE are congruent alternate interior angles formed by the parallel segments. So m∠DFB = m∠FDE = 54°.

The correct choice is **(2)**.

**11.** If a parallelogram's diagonals are perpendicular or if the diagonals bisect the angles, the parallelogram must be a rhombus. So statements I and II always describe a rhombus. If a parallelogram's diagonals form 4 congruent isosceles triangles as shown in the figure, then all 4 sides of the parallelogram must be congruent and the diagonals must be congruent. This is a square, which is also a rhombus. Statement III also always describes a rhombus.

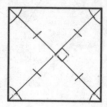

The correct choice is (**4**).

**12.** Use the completing the square procedure to rewrite the equation in the form $(x - h)^2 + (y - k)^2 = r^2$. First group the $x$-terms and $y$-terms on the left. Then move the constant term to the right:

$$x^2 + y^2 - 12y + 20 = 0$$
$$x^2 + y^2 - 12y \phantom{+ 20} = -20$$

Next find the constant terms needed to complete the square:

$$\text{constant for the } y\text{-terms} = \left(\frac{1}{2} \cdot \text{coefficient of } y\right)^2$$
$$= \left(\frac{1}{2} \cdot (-12)\right)^2$$
$$= (-6)^2$$
$$= 36$$

Add the required constant to each side of the equation and factor:

$$x^2 + y^2 - 12y + 36 = -20 + 36$$
$$x^2 + (y - 6)^2 = 16$$

The values of $h$ and $k$ are 0 and 6. Be careful with the signs because $h$ and $k$ are the values subtracted from $x$ and $y$. The center is located at $(0, 6)$. Since $r^2 = 16$, the radius is equal to 4.

The correct choice is (**1**).

**13.** Set up an appropriate trig ratio, and then use the inverse function to find $m\angle S$. Relative to $\angle S$, $ST$ is the adjacent and $RS$ is the hypotenuse. So use cosine.

$$\cos S = \frac{\text{adj}}{\text{hyp}}$$

$$= \frac{ST}{RS}$$

$$= \frac{60}{65}$$

$$m\angle S = \cos^{-1}\frac{60}{65}$$

$$= 23°$$

The correct choice is (**1**).

**14.** Making a sketch of the situation is always a good idea. The figure below shows the image of $\triangle ABC$ after a dilation and translation. Check each of the statements.

- Statement I: Since a dilation changes the size of a figure (unless the scale factor is 1), $\triangle ABC$ cannot be congruent to $\triangle A'B'C'$. Statement I is false.

- Statement II: A dilation always results in an image similar to the pre-image, so $\triangle ABC \sim \triangle A'B'C'$. Statement II is true.

- Statement III: Because of a discrepancy in wording, this statement could be either true or false. For example, most translations will cause the corresponding segments to be parallel. Therefore, $\overline{AB}$ // $\overline{A'B'}$. However, if the translation caused $\overline{AB}$ and $\overline{A'B'}$ to be on the same line, the line segments would not be parallel. Statement III could be either true or false.

- Statement IV: Look at the drawing shown below. As long as the scale factor of the dilation is not 1, $AA'$ is not parallel to $BB'$. Statement IV is false.

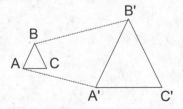

Both choice (**1**) and choice (**3**) were accepted as correct.

**15.** Find the $x$- and $y$-coordinates of $P$ using the following formulas:

$$\text{ratio} = \frac{x - x_1}{x_2 - x}$$

$$\text{ratio} = \frac{y - y_1}{y_2 - y}$$

When using $R(-4, 5)$ and $W(6, 20)$, $x_1 = -4$, $x_2 = 6$, $y_1 = 5$, $y_2 = 20$, and the ratio $= \frac{2}{3}$.

$$\text{ratio} = \frac{x - x_1}{x_2 - x} \qquad\qquad \text{ratio} = \frac{y - y_1}{y_2 - y}$$

$$\frac{2}{3} = \frac{x - (-4)}{6 - x} \qquad\qquad \frac{2}{3} = \frac{y - 5}{20 - y}$$

$$3(x + 4) = 2(6 - x) \qquad\qquad 3(y - 5) = 2(20 - y)$$

$$3x + 12 = 12 - 2x \qquad\qquad 3y - 15 = 40 - 2y$$

$$5x + 12 = 12 \qquad\qquad 5y - 15 = 40$$

$$5x = 0 \qquad\qquad 5y = 55$$

$$x = 0 \qquad\qquad y = 11$$

The coordinates of P are $(0, 11)$. The correct choice is **(2)**.

**16.** The formula for the volume of a pyramid, $V = \frac{1}{3}Bh$, can be found on the reference sheet. In this formula, $B$ represents the area of the square base. By using the formula area $= s^2$ for a square, we can replace $B$ with $s^2$:

$$V = \frac{1}{3}s^2 h$$

$$84 = \frac{1}{3}(s^2)(7)$$

$$252 = s^2 (7)$$

$$36 = s^2$$

$$s = 6$$

The correct choice is **(1)**.

**17.** In this problem, label all known angles. Then begin finding the measures of the other angles one by one until you get the desired angle.

$m\angle TMO = m\angle TMR = 28°$     $\overline{MS}$ is an angle bisector

$m\angle NMO = m\angle TMO + m\angle TMR$     angle sum theorem

$m\angle MON + m\angle N + m\angle NMO = 180°$     angle sum theorem

$m\angle MON + 40° + 56° = 180°$

$m\angle MON = 84°$

$m\angle MOT = \frac{1}{2}m\angle MON$     $\overline{OR}$ is an angle bisector

$= \frac{1}{2} \cdot 84°$

$= 42°$

$m\angle MTO + m\angle TMO + m\angle MOT = 180°$     angle sum theorem in $\triangle MOT$

$m\angle MTO + 28° + 42° = 180°$

$m\angle MTO = 110°$

$m\angle OTS + m\angle MTO = 180°$     linear pair sums to 180°

$m\angle OTS + 110° = 180°$

$m\angle OTS = 70°$

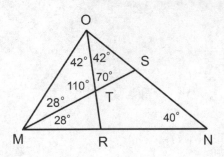

The correct choice is **(4)**.

18. Rotating $\triangle ABC$ about $\overline{AB}$ results in a cone with height $AB$ and radius $AC$. Apply the volume formula for a cone:

$$V = \frac{1}{3}\pi r^2 h$$

$$= \frac{1}{3}\pi(4^2)(6)$$

$$= 32\pi$$

The correct choice is (**1**).

19. Perpendicular lines have slopes that are negative reciprocals of each other. So first find the slope of the given line.

$$2y = 3x - 10$$

$$y = \frac{3}{2}x - 5$$

The slope of the given line is $\frac{3}{2}$, so the slope of the perpendicular line is $-\frac{2}{3}$. Use this slope and the given point $(-6, 1)$ to find the $y$-intercept:

$$y = mx + b$$
$$1 = -\frac{2}{3}(-6) + b$$
$$1 = 4 + b$$
$$b = -3$$

Using the new slope and intercept, we find the equation of the line is $y = -\frac{2}{3}x - 3$.

The correct choice is (**2**).

20. One pair of opposite sides of $BLUE$ are congruent. To prove quadrilateral $BLUE$ is a parallelogram, we need to show either that the same pair of sides are also parallel or that the other pair of opposite sides are congruent. Choice (2) states that the same pair of sides are also parallel.

The correct choice is (**2**).

**21.** The ladder, wall, and ground form a right triangle as shown in the figure. Use an appropriate trig ratio to find the height, $x$. Relative to the $70°$ angle, the hypotenuse is 20 and the opposite is $x$. Opposite and hypotenuse indicate the sine ratio:

$$\sin 71° = \frac{\text{opp}}{\text{hyp}}$$

$$\sin 71° = \frac{x}{20}$$

$$x = 20 \sin 71°$$

$$= 18.91$$

When rounded to the *nearest foot*, $x = 19$.

The correct choice is (**4**).

**22.** This question was invalidated by the New York State Education Department. Credit was awarded for any of the choices.

**23.** The diameter of the cylinder is 2.5, so its radius is 1.25. From the figure, we can find the height of the prism:

$$\text{prism height} = \text{total height} - \text{cylinder radius}$$

$$= 2.5 - 1.25$$

$$= 1.25$$

We also need to convert the length of 27 feet into inches:

$$27 \text{ ft} \cdot \frac{12 \text{ in}}{\text{ft}} = 324 \text{ in}$$

The total volume is found by adding the volume of the prism and the volume of the half-cylinder:

$$\text{volume} = V_{\text{prism}} + V_{\text{half-cylinder}}$$

$$\text{volume} = lwh + \frac{1}{2}\pi r^2 h$$

$$= 324 \cdot 2.5 \cdot 1.25 + \frac{1}{2}\pi(1.25)^2(324)$$

$$= 1012.5 + 795.2156$$

$$= 1807.7156$$

To the *nearest cubic inch*, the railing will contain 1808 cubic inches of metal.

The correct choice is **(3)**.

**24.** This question was invalidated by the New York State Education Department. Credit was awarded for any of the choices.

## PART II

**25.** Use the construction for a perpendicular line through a point off the line. Start by extending $\overline{ML}$ and making an arc from point $J$ that crosses the extended line in two places. See figure (a). Next, from those intersection points make a pair of arcs that intersect below the figure, shown in figure (b). Connect point $J$ to the intersecting arcs to form a perpendicular line. Let point $N$ be where the perpendicular line intersects $\overline{ML}$. $\overline{JN}$ is the altitude.

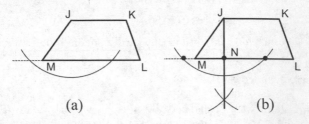

(a)                   (b)

**26.** Use the formula for the area of a sector. Remember that $\theta$ is the measure of the central angle in degrees.

$$A_{sector} = \pi r^2 \frac{\theta}{360}$$

$$A_{sector} = \pi (4.5)^2 \frac{40}{360}$$

$$= \mathbf{2.25\pi} \text{ or } \frac{9}{4}\pi$$

**27.** Cavalieri's principle states that if the cross sections of two solids always have the same area and if their heights are equal, then the volumes of the solids are equal. The area of each base is $\frac{1}{2}(8)(5) = 20$ and the height of each is 14. So the volumes of figure $A$ and figure $B$ are equal.

**28.** Calculate the radius of each volleyball using the formula for the volume of a sphere.

Partially inflated:          Fully inflated:

$$V_{sphere} = \frac{4}{3}\pi r^3 \qquad\qquad V_{sphere} = \frac{4}{3}\pi r^3$$

$$180 = \frac{4}{3}\pi r^3 \qquad\qquad 294 = \frac{4}{3}\pi r^3$$

$$r^3 = 42.9718 \qquad\qquad r^3 = 70.1873$$

$$r = \sqrt[3]{42.9718} = 3.502 \qquad\qquad r = \sqrt[3]{70.1873} = 4.124$$

The difference in the radii is $4.124 - 3.502 = 0.622$ inches. Rounded to the *nearest tenth of an inch*, the difference is **0.6 inches**.

**29.** Altitude $\overline{CD}$ is perpendicular to side $\overline{AB}$, so $\angle ADC$ must be a right angle. $\angle ACB$ is also a right angle because it is opposite the hypotenuse, therefore $\angle ADC \cong \angle ACB$. The two triangles share $\angle A$, which is congruent to itself. Since there are two pairs of corresponding congruent angles, we know $\triangle ABC \sim \triangle ACD$ by the AA postulate.

**30.** A congruence statement for the two triangles is $\triangle ABC \cong \triangle DEF$. One possible transformation would be to map the first vertex by translating. Then map the others by rotating. So translate $\triangle ABC$ along vector $\overrightarrow{AD}$ so point $A$ maps to point $D$. Then rotate $\triangle ABC$ about point $D$ so point $C$ maps to point $F$. This will cause point $B$ to map to point $E$.

**31.** The key to this problem is to realize that the point $(4, 2)$ lies on the line to be dilated. The point satisfies the equation of the line, which can be checked either by graphing the point and line or by substitution. By using substitution, we find the following:

$$3x + 4y = 20$$
$$3(4) + 4(2) = 20$$
$$20 = 20$$

A dilation of a line through a point on the line does not change the line. Therefore the image is the original line, $3x + 4y = 20$.

## PART III

**32.** The graph is shown in the accompanying figure. Be sure to include all labels.

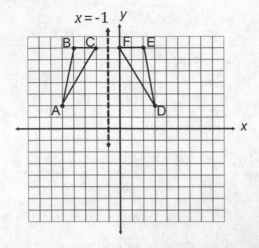

The single transformation that maps $\triangle ABC$ to $\triangle DEF$ is a reflection over the vertical line $x = -1$. This transformation results in $\triangle ABC \cong \triangle DEF$ because reflections are rigid motions that preserve the size and shape of the preimage.

**33.** We can prove the two segments parallel by first proving the triangles are congruent and then using CPCTC to show the alternate interior angles are congruent.

| Statements | Reasons |
|---|---|
| 1. $\overline{RS}$ and $\overline{TV}$ bisect each other at $X$ | 1. Given |
| 2. $X$ is the midpoint of $\overline{RS}$ and $\overline{TV}$ | 2. A bisector intersects a segment at the midpoint |
| 3. $\overline{XT} \cong VX, RX \cong XS$ | 3. The midpoint divides a segment into two congruent segments |
| 4. $\angle RXT \cong \angle SXV$ | 4. Vertical angles are congruent |
| 5. $\triangle RXT \cong SXV$ | 5. SAS |
| 6. $\angle T \cong \angle V$ | 6. CPCTC |
| 7. $\overline{TR} /\!/ \overline{SV}$ | 7. Two segments are parallel if the alternate interior angles formed by a transversal are congruent |

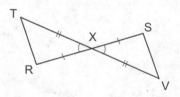

**34.** From the given figure, we see that the length of the pole must be 1 foot greater than the diameter of the tank. Begin by converting the volume from gallons to cubic feet. Then use the volume formula to calculate the radius of the tank:

$$V = 20{,}000 \text{ gallons} \cdot \frac{1 \text{ ft}^3}{7.48 \text{ gallons}} \qquad \text{convert gallons to ft}^3$$

$$= 2673.7967 \text{ ft}^3$$

$$V = \pi r^2 h \qquad\qquad\qquad\qquad\qquad \text{volume formula for cylinder}$$

$$2673.7967 = \pi r^2 \cdot 34.5$$

$$24.6694 = r^2$$

$$r = \sqrt{24.6694}$$

$$= 4.9668 \text{ ft}$$

$$\text{diameter} = 2r \qquad\qquad\qquad\qquad \text{diameter = twice the radius}$$

$$= 9.9336$$

$$\text{pole length} = \text{diameter} + 1 \text{ ft} \qquad\qquad \text{add 1 foot to the diameter}$$

$$= 10.9336$$

Rounded to the *nearest tenth of a foot*, the pole must be **10.9** feet long.

## PART IV

**35.** One method of approaching this proof is to calculate the slopes of each side and each diagonal. Use the slope formula to show that opposite sides are parallel and that the diagonals are perpendicular:

$$\text{slope} = \frac{y_2 - y_1}{x_2 - x_1}$$

Slopes of the sides:

slope of $\overline{PQ} = \dfrac{8-3}{3-(-2)} = \dfrac{5}{5} = 1$

slope of $\overline{RS} = \dfrac{-4-1}{-1-4} = \dfrac{-5}{-5} = 1$

slope of $\overline{QR} = \dfrac{1-8}{4-3} = \dfrac{-7}{1} = -7$

slope of $\overline{SP} = \dfrac{-4-3}{-1-(-2)} = \dfrac{-7}{1} = -7$

Slopes of the diagonals:

slope of $\overline{PR} = \dfrac{1-3}{4-(-2)} = \dfrac{-2}{6} = -\dfrac{1}{3}$

slope of $\overline{SQ} = \dfrac{-4-8}{-1-3} = \dfrac{-12}{-4} = 3$

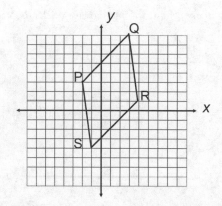

The slopes of opposite sides are equal, making them parallel. The slopes of the diagonals are negative reciprocals, making them perpendicular. Therefore $PQRS$ is a rhombus.

The slopes of consecutive sides are 1 and $-7$. Since they are not negative reciprocals, the consecutive sides are not perpendicular. Therefore $PQRS$ is not a square.

36. The airplane, its altitude, and the ground form a right triangle. The triangle in figure (a) shows the 15° angle of elevation when Freda first detects the airplane. It is a distance $x$ away from Freda. Figure (b) shows the 52° angle of elevation and a distance $y$.

Use trig ratios to calculate $x$ and $y$. Relative to the angle of elevation, the 6250 altitude is the opposite. The distances $x$ and $y$ are adjacents. The tangent ratio applies for both figures.

$$\tan 15° = \frac{6250}{x} \qquad\qquad \tan 52° = \frac{6250}{y}$$

$$x \tan 15° = 6250 \qquad\qquad y \tan 52° = 6250$$

$$x = \frac{6250}{\tan 15°} \qquad\qquad y = \frac{6250}{\tan 52°}$$

$$x = 23{,}325.3175 \text{ ft} \qquad\qquad y = 4883.0351 \text{ ft}$$

distance traveled $= x - y$

$\qquad\qquad\quad = 23{,}325.3175 \text{ ft} - 4883.0351 \text{ ft}$

$\qquad\qquad\quad = 18{,}442.2824 \text{ ft}$

Rounded to the *nearest foot*, the plane traveled **18,442 feet**.

To calculate the speed in miles per hour, we need to convert the distance from feet to miles and the time of 1 minute to hours. The conversion for feet to miles can be found on the reference table.

$$18,442 \text{ ft} \cdot \frac{1 \text{ mile}}{5280 \text{ ft}} = 3.492803 \text{ miles}$$

$$1 \text{ minute} \cdot \frac{1 \text{ hour}}{60 \text{ minutes}} = 0.016666 \text{ hours}$$

$$\text{speed} = \frac{\text{distance}}{\text{time}}$$

$$= \frac{3.492803 \text{ miles}}{0.016666 \text{ hours}}$$

$$= 209.576 \frac{\text{miles}}{\text{hour}}$$

(a)                    (b)

The speed of the airplane to the *nearest mile per hour* is **210 miles per hour**.

| Topic | Question Numbers | Number of Points | Your Points | Your Percentage |
|---|---|---|---|---|
| 1. Basic Angle and Segment Relationships | | | | |
| 2. Angle and Segment Relationships in Triangles and Polygons | 10, 17 | 2 + 2 = 4 | | |
| 3. Constructions | 25 | 2 | | |
| 4. Transformations | 1, 6, 7, 14, 22, 30, 31, 32 | 2 + 2 + 2 + 2 + 2 + 2 + 2 + 4 = 18 | | |
| 5. Triangle Congruence | 9, 33 | 2 + 4 = 6 | | |
| 6. Lines, Segments, and Circles on the Coordinate Plane | 2, 12, 15, 19 | 2 + 2 + 2 + 2 = 8 | | |
| 7. Similarity | 5, 24, 29 | 2 + 2 + 2 = 6 | | |
| 8. Trigonometry | 3, 13, 21, 36 | 2 + 2 + 2 + 6 = 12 | | |
| 9. Parallelograms | 11, 20 | 2 + 2 = 4 | | |
| 10. Coordinate Geometry Proofs | 35 | 6 | | |
| 11. Volume and Solids | 16, 18, 27, 28 | 2 + 2 + 2 + 2 = 8 | | |
| 12. Modeling | 23, 34 | 2 + 4 = 6 | | |
| 13. Circles | 4, 8, 26 | 2 + 2 + 2 = 6 | | |

# HOW TO CONVERT YOUR RAW SCORE TO YOUR GEOMETRY REGENTS EXAMINATION SCORE

The conversion chart below can be used to determine your final score on the June 2017 Regents Exam in Geometry. To find your final exam score, locate in the "Raw Score" column the total number of points you scored out of a possible 86. Then locate in the adjacent column to the right the scale score that corresponds to your raw score. The scale score is your final score.

| Raw Score | Scale Score | Performance Level | Raw Score | Scale Score | Performance Level | Raw Score | Scale Score | Performance Level |
|---|---|---|---|---|---|---|---|---|
| 86 | 100 | 5 | 57 | 79 | 3 | 28 | 59 | 2 |
| 85 | 99 | 5 | 56 | 78 | 3 | 27 | 58 | 2 |
| 84 | 97 | 5 | 55 | 78 | 3 | 26 | 57 | 2 |
| 83 | 96 | 5 | 54 | 77 | 3 | 25 | 55 | 2 |
| 82 | 95 | 5 | 53 | 77 | 3 | 24 | 54 | 1 |
| 81 | 94 | 5 | 52 | 77 | 3 | 23 | 53 | 1 |
| 80 | 93 | 5 | 51 | 76 | 3 | 22 | 51 | 1 |
| 79 | 92 | 5 | 50 | 76 | 3 | 21 | 49 | 1 |
| 78 | 91 | 5 | 49 | 75 | 3 | 20 | 48 | 1 |
| 77 | 91 | 5 | 48 | 75 | 3 | 19 | 46 | 1 |
| 76 | 90 | 5 | 47 | 74 | 3 | 18 | 44 | 1 |
| 75 | 89 | 5 | 46 | 74 | 3 | 17 | 43 | 1 |
| 74 | 88 | 5 | 45 | 73 | 3 | 16 | 41 | 1 |
| 73 | 87 | 5 | 44 | 72 | 3 | 15 | 39 | 1 |
| 72 | 87 | 5 | 43 | 72 | 3 | 14 | 37 | 1 |
| 71 | 86 | 5 | 42 | 71 | 3 | 13 | 35 | 1 |
| 70 | 86 | 5 | 41 | 71 | 3 | 12 | 32 | 1 |
| 69 | 85 | 5 | 40 | 70 | 3 | 11 | 30 | 1 |
| 68 | 84 | 4 | 39 | 69 | 3 | 10 | 28 | 1 |
| 67 | 84 | 4 | 38 | 69 | 3 | 9 | 25 | 1 |
| 66 | 83 | 4 | 37 | 68 | 3 | 8 | 23 | 1 |
| 65 | 83 | 4 | 36 | 67 | 3 | 7 | 20 | 1 |
| 64 | 82 | 4 | 35 | 66 | 3 | 6 | 18 | 1 |
| 63 | 82 | 4 | 34 | 65 | 3 | 5 | 15 | 1 |
| 62 | 81 | 4 | 33 | 64 | 2 | 4 | 12 | 1 |
| 61 | 81 | 4 | 32 | 63 | 2 | 3 | 9 | 1 |
| 60 | 80 | 4 | 31 | 62 | 2 | 2 | 6 | 1 |
| 59 | 80 | 4 | 30 | 61 | 2 | 1 | 3 | 1 |
| 58 | 79 | 3 | 29 | 60 | 2 | 0 | 0 | 1 |

# Examination August 2017
## Geometry (Common Core)

## GEOMETRY REFERENCE SHEET

1 inch = 2.54 centimeters
1 meter = 39.37 inches
1 mile = 5280 feet
1 mile = 1760 yards
1 mile = 1.609 kilometers
1 kilometer = 0.62 mile
1 pound = 16 ounces
1 pound = 0.454 kilogram
1 kilogram = 2.2 pounds

1 ton = 2000 pounds
1 cup = 8 fluid ounces
1 pint = 2 cups
1 quart = 2 pints
1 gallon = 4 quarts
1 gallon = 3.785 liters
1 liter = 0.264 gallon
1 liter = 1000 cubic centimeters

| Triangle | $A = \frac{1}{2}bh$ |
| --- | --- |
| Parallelogram | $A = bh$ |
| Circle | $A = \pi r^2$ |
| Circle | $C = \pi d$ or $C = 2\pi r$ |
| General Prisms | $V = Bh$ |
| Cylinder | $V = \pi r^2 h$ |
| Sphere | $V = \frac{4}{3}\pi r^3$ |

| Cone | $V = \dfrac{1}{3}\pi r^2 h$ |
|---|---|
| Pyramid | $V = \dfrac{1}{3}Bh$ |
| Pythagorean Theorem | $a^2 + b^2 = c^2$ |
| Quadratic Formula | $x = \dfrac{-b \pm \sqrt{b^2 - 4ac}}{2a}$ |
| Arithmetic Sequence | $a_n = a_1 + (n-1)d$ |
| Geometric Sequence | $a_n = a_1 r^{n-1}$ |
| Geometric Series | $S_n = \dfrac{a_1 - a_1 r^n}{1-r}$ where $r \neq 1$ |
| Radians | $1$ radian $= \dfrac{180}{\pi}$ degrees |
| Degrees | $1$ degree $= \dfrac{\pi}{180}$ radians |
| Exponential Growth/Decay | $A = A_0 e^{k(t-t_0)} + B_0$ |

## PART I

Answer all 24 questions in this part. Each correct answer will receive 2 credits. No partial credit will be allowed. For each statement or question, write in the space provided the numeral preceding the word or expression that best completes the statement or answers the question. [48 credits]

1 A two-dimensional cross section is taken of a three-dimensional object. If this cross section is a triangle, what can *not* be the three-dimensional object?

(1) cone          (3) pyramid

(2) cylinder      (4) rectangular prism     1 _____

2 The image of $\triangle DEF$ is $\triangle D'E'F'$. Under which transformation will the triangles *not* be congruent?

(1) a reflection through the origin

(2) a reflection over the line $y = x$

(3) a dilation with a scale factor of 1 centered at $(2, 3)$

(4) a dilation with a scale factor of $\frac{3}{2}$ centered at the

     origin                                    2 _____

3 The vertices of square $RSTV$ have coordinates $R(-1, 5)$, $S(-3, 1)$, $T(-7, 3)$, and $V(-5, 7)$. What is the perimeter of $RSTV$?

(1) $\sqrt{20}$          (3) $4\sqrt{20}$

(2) $\sqrt{40}$          (4) $4\sqrt{40}$         3 _____

**4** In the diagram below of circle $O$, chord $\overline{CD}$ is parallel to diameter $\overline{AOB}$ and $\text{m}\overset{\frown}{CD} = 130$.

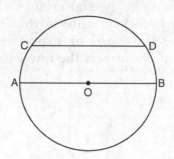

What is $\text{m}\overset{\frown}{AC}$?

(1) 25             (3) 65

(2) 50             (4) 115          4 _____

**5** In the diagram below, $\overline{AD}$ intersects $\overline{BE}$ at $C$, and $\overline{AB} \parallel \overline{DE}$.

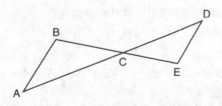

If $CD = 6.6$ cm, $DE = 3.4$ cm, $CE = 4.2$ cm, and $BC = 5.25$ cm, what is the length of $\overline{AC}$, to the *nearest hundredth of a centimeter*?

(1) 2.70            (3) 5.28

(2) 3.34            (4) 8.25          5 _____

**6** As shown in the graph below, the quadrilateral is a rectangle.

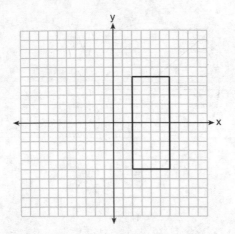

Which transformation would *not* map the rectangle onto itself?

(1) a reflection over the $x$-axis
(2) a reflection over the line $x = 4$
(3) a rotation of 180° about the origin
(4) a rotation of 180° about the point $(4, 0)$          6 _____

**7** In the diagram below, triangle $ACD$ has points $B$ and $E$ on sides $\overline{AC}$ and $\overline{AD}$, respectively, such that $\overline{BE} \parallel \overline{CD}$, $AB = 1$, $BC = 3.5$, and $AD = 18$.

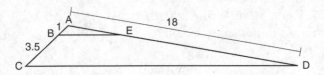

What is the length of $\overline{AE}$, to the *nearest tenth*?

(1) 14.0          (3) 3.3
(2) 5.1          (4) 4.0          7 _____

**8** In the diagram below of parallelogram ROCK, m∠C is 70° and m∠ROS is 65°.

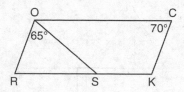

What is m∠KSO?

(1) 45°  (3) 115°
(2) 110°  (4) 135°

8 _____

**9** In the diagram below, ∠GRS ≅ ∠ART, GR = 36, SR = 45, AR = 15, and RT = 18.

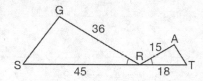

Which triangle similarity statement is correct?

(1) △GRS ~ △ART by AA.
(2) △GRS ~ △ART by SAS.
(3) △GRS ~ △ART by SSS.
(4) △GRS is not similar to △ART.

9 _____

**10** The line represented by the equation $4y = 3x + 7$ is transformed by a dilation centered at the origin. Which linear equation could represent its image?

(1) $3x - 4y = 9$  (3) $4x - 3y = 9$
(2) $3x + 4y = 9$  (4) $4x + 3y = 9$

10 _____

**11** Given $\triangle ABC$ with m$\angle B = 62°$ and side $\overline{AC}$ extended to $D$, as shown below.

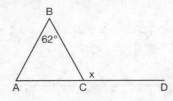

What value of $x$ makes $\overline{AB} \cong \overline{CB}$?

(1) 59°  (3) 118°

(2) 62°  (4) 121°

11 _____

**12** In the diagram shown below, $\overline{PA}$ is tangent to circle $T$ at $A$, and secant $\overline{PBC}$ is drawn where point $B$ is on circle $T$.

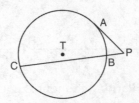

If $PB = 3$ and $BC = 15$, what is the length of $\overline{PA}$?

(1) $3\sqrt{5}$  (3) 3

(2) $3\sqrt{6}$  (4) 9

12 _____

**13** A rectangle whose length and width are 10 and 6, respectively, is shown below. The rectangle is continuously rotated around a straight line to form an object whose volume is $150\pi$.

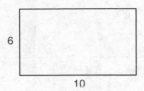

Which line could the rectangle be rotated around?

(1) a long side
(2) a short side
(3) the vertical line of symmetry
(4) the horizontal line of symmetry      13 _____

**14** If *ABCD* is a parallelogram, which statement would prove that *ABCD* is a rhombus?

(1) $\angle ABC \cong \angle CDA$      (3) $\overline{AC} \perp \overline{BD}$
(2) $\overline{AC} \cong \overline{BD}$      (4) $\overline{AB} \perp \overline{CD}$      14 _____

**15** To build a handicapped-access ramp, the building code states that for every 1 inch of vertical rise in height, the ramp must extend out 12 inches horizontally, as shown in the diagram below.

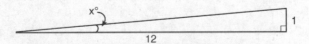

What is the angle of inclination, *x*, of this ramp, to the *nearest hundredth of a degree*?

(1) 4.76      (3) 85.22
(2) 4.78      (4) 85.24      15 _____

**16** In the diagram below of $\triangle ABC$, $D$, $E$, and $F$ are the midpoints of $\overline{AB}$, $\overline{BC}$, and $\overline{CA}$, respectively.

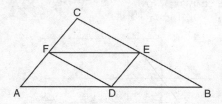

What is the ratio of the area of $\triangle CFE$ to the area of $\triangle CAB$?

(1) 1 : 1                    (3) 1 : 3
(2) 1 : 2                    (4) 1 : 4                    16 _____

**17** The coordinates of the endpoints of $\overline{AB}$ are $A(-8, -2)$ and $B(16, 6)$. Point $P$ is on $\overline{AB}$. What are the coordinates of point $P$, such that $AP : PB$ is 3 : 5?

(1) (1, 1)                    (3) (9.6, 3.6)
(2) (7, 3)                    (4) (6.4, 2.8)                    17 _____

**18** Kirstie is testing values that would make triangle *KLM* a right triangle when $\overline{LN}$ is an altitude, and *KM* = 16, as shown below.

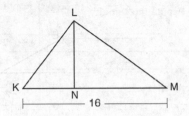

Which lengths would make triangle *KLM* a right triangle?

(1) *LM* = 13 and *KN* = 6
(2) *LM* = 12 and *NM* = 9
(3) *KL* = 11 and *KN* = 7
(4) *LN* = 8 and *NM* = 10          18 _____

**19** In right triangle ABC, m∠*A* = 32°, m∠*B* = 90°, and *AC* = 6.2 cm. What is the length of $\overline{BC}$, to the *nearest tenth of a centimeter*?

(1) 3.3          (3) 5.3
(2) 3.9          (4) 11.7          19 _____

**20** The 2010 U.S. Census populations and population densities are shown in the table below.

| State | Population Density $\left(\frac{\text{people}}{\text{mi}^2}\right)$ | Population in 2010 |
|---|---|---|
| Florida | 350.6 | 18,801,310 |
| Illinois | 231.1 | 12,830,632 |
| New York | 411.2 | 19,378,102 |
| Pennsylvania | 283.9 | 12,702,379 |

Based on the table above, which list has the states' areas, in square miles, in order from largest to smallest?

(1) Illinois, Florida, New York, Pennsylvania
(2) New York, Florida, Illinois, Pennsylvania
(3) New York, Florida, Pennsylvania, Illinois
(4) Pennsylvania, New York, Florida, Illinois      20 _____

**21** In a right triangle, $\sin(40 - x)° = \cos(3x)°$. What is the value of $x$?

(1) 10          (3) 20
(2) 15          (4) 25      21 _____

**22** A regular decagon is rotated $n$ degrees about its center, carrying the decagon onto itself. The value of $n$ could be

(1) 10°          (3) 225°
(2) 150°          (4) 252°      22 _____

**23** In a circle with a diameter of 32, the area of a sector is $\dfrac{512\pi}{3}$. The measure of the angle of the sector, in radians, is

(1) $\dfrac{\pi}{3}$         (3) $\dfrac{16\pi}{3}$

(2) $\dfrac{4\pi}{3}$       (4) $\dfrac{64\pi}{3}$        23 \_\_\_\_\_

**24** What is an equation of the perpendicular bisector of the line segment shown in the diagram below?

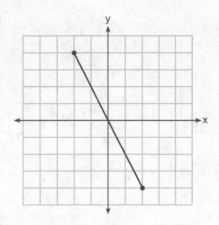

(1) $y + 2x = 0$       (3) $2y + x = 0$

(2) $y - 2x = 0$       (4) $2y - x = 0$        24 \_\_\_\_\_

## PART II

Answer all 7 questions in this part. Each correct answer will receive 2 credits. Clearly indicate the necessary steps, including appropriate formula substitutions, diagrams, graphs, charts, etc. For all questions in this part, a correct numerical answer with no work shown will receive only 1 credit. [14 credits]

**25** Sue believes that the two cylinders shown in the diagram below have equal volumes.

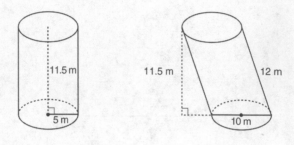

Is Sue correct? Explain why.

**26** In the diagram of rhombus *PQRS* below, the diagonals $\overline{PR}$ and $\overline{QS}$ intersect at point *T*, *PR* = 16, and *QS* = 30. Determine and state the perimeter of *PQRS*.

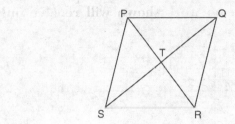

27 Quadrilateral *MATH* and its image *M"A"T"H"* are graphed on the set of axes below.

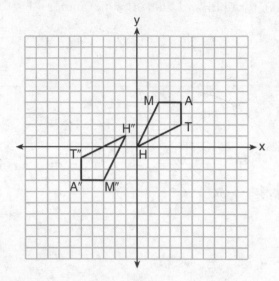

Describe a sequence of transformations that maps quadrilateral *MATH* onto quadrilateral *M"A"T"H"*.

**28** Using a compass and straightedge, construct a regular hexagon inscribed in circle $O$. [Leave all construction marks.]

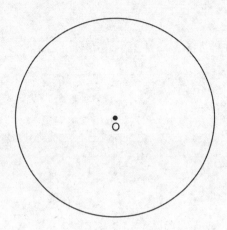

**29** The coordinates of the endpoints of $\overline{AB}$ are $A(2, 3)$ and $B(5, -1)$. Determine the length of $\overline{A'B'}$, the image of $\overline{AB}$, after a dilation of $\frac{1}{2}$ centered at the origin.

[The use of the set of axes below is optional.]

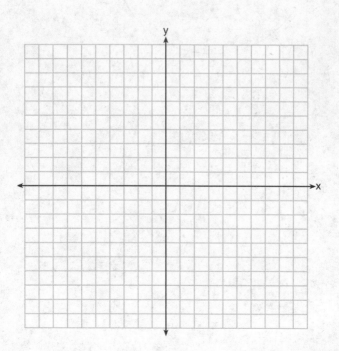

**30** In the diagram below of △ABC and △XYZ, a sequence of rigid motions maps ∠A onto ∠X, ∠C onto ∠Z, and $\overline{AC}$ onto $\overline{XZ}$.

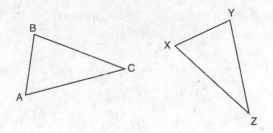

Determine and state whether $\overline{BC} \cong \overline{YZ}$. Explain why.

**31** Determine and state the coordinates of the center and the length of the radius of a circle whose equation is $x^2 + y^2 - 6x = 56 - 8y$.

## PART III

**Answer all 3 questions in this part. Each correct answer will receive 4 credits. Clearly indicate the necessary steps, including appropriate formula substitutions, diagrams, graphs, charts, etc. For all questions in this part, a correct numerical answer with no work shown will receive only 1 credit.** [12 credits]

**32** Triangle $PQR$ has vertices $P(-3, -1)$, $Q(-1, 7)$, and $R(3, 3)$, and points $A$ and $B$ are midpoints of $\overline{PQ}$ and $\overline{RQ}$, respectively. Use coordinate geometry to prove that $\overline{AB}$ is parallel to $\overline{PR}$ and is half the length of $\overline{PR}$.

[The use of the set of axes below is optional.]

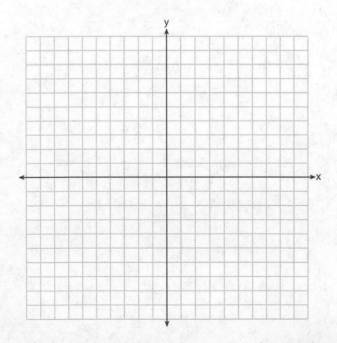

**33** In the diagram below of circle $O$, tangent $\overleftrightarrow{EC}$ is drawn to diameter $\overline{AC}$. Chord $\overline{BC}$ is parallel to secant $\overleftrightarrow{ADE}$, and chord $\overline{AB}$ is drawn.

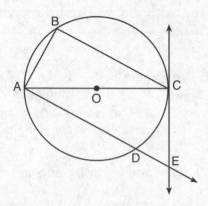

Prove: $\dfrac{BC}{CA} = \dfrac{AB}{EC}$

**34** Keira has a square poster that she is framing and placing on her wall. The poster has a diagonal 58 cm long and fits exactly inside the frame. The width of the frame around the picture is 4 cm.

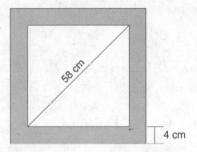

4 cm

Determine and state the total area of the poster and frame to the *nearest tenth of a square centimeter*.

## PART IV

Answer the 2 questions in this part. Each correct answer will receive 6 credits. Clearly indicate the necessary steps, including appropriate formula substitutions, diagrams, graphs, charts, etc. For all questions in this part, a correct numerical answer with no work shown will receive only 1 credit. [12 credits]

35 Isosceles trapezoid $ABCD$ has bases $\overline{DC}$ and $\overline{AB}$ with nonparallel legs $\overline{AD}$ and $\overline{BC}$. Segments $AE$, $BE$, $CE$, and $DE$ are drawn in trapezoid $ABCD$ such that $\angle CDE \cong \angle DCE$, $\overline{AE} \perp \overline{DE}$, and $\overline{BE} \perp \overline{CE}$.

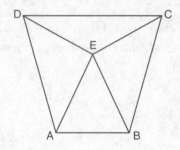

Prove $\triangle ADE \cong \triangle BCE$ and prove $\triangle AEB$ is an isosceles triangle.

**36** A rectangular in-ground pool is modeled by the prism below. The inside of the pool is 16 feet wide and 35 feet long. The pool has a shallow end and a deep end, with a sloped floor connecting the two ends. Without water, the shallow end is 9 feet long and 4.5 feet deep, and the deep end of the pool is 12.5 feet long.

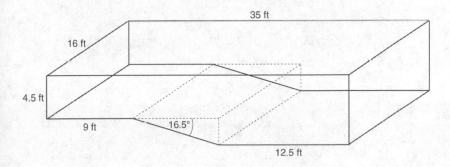

If the sloped floor has an angle of depression of 16.5 degrees, what is the depth of the pool at the deep end, to the *nearest tenth of a foot*?

Find the volume of the inside of the pool to the *nearest cubic foot*.

**Question 36 is continued on the next page.**

## Question 36 continued

A garden hose is used to fill the pool. Water comes out of the hose at a rate of 10.5 gallons per minute. How much time, to the *nearest hour*, will it take to fill the pool 6 inches from the top?
[1 ft$^3$ = 7.48 gallons]

# Answers
# August 2017
## Geometry (Common Core)

## Answer Key

### PART I

| | | | | | |
|---|---|---|---|---|---|
| **1.** (2) | **5.** (4) | **9.** (4) | **13.** (3) | **17.** (1) | **21.** (4) |
| **2.** (4) | **6.** (3) | **10.** (1) | **14.** (3) | **18.** (2) | **22.** (4) |
| **3.** (3) | **7.** (4) | **11.** (4) | **15.** (1) | **19.** (1) | **23.** (2) |
| **4.** (1) | **8.** (4) | **12.** (2) | **16.** (4) | **20.** (1) | **24.** (4) |

### PART II

**25.** The heights and cross-sectional areas are equal, so the volumes must be equal.

**26.** 68

**27.** A rotation of $180°$ about the origin followed by a translation of 1 unit up and 1 unit left

**28.** See detailed solution for the construction.

**29.** $\dfrac{5}{2}$

**30.** Rigid motions preserve angle measure and length, so the triangles are congruent and $\overline{BC} \cong \overline{YZ}$ by CPCTC.

**31.** Center $(3, -4)$   radius = 3

### PART III

**32.** Slopes of $\overline{AB}$ and $\overline{PR}$ are both $\dfrac{2}{3}$.
$AB = \sqrt{13}$ and $\overline{PR} = 2\sqrt{13}$.

**33.** Prove $\triangle ABC$ is similar to $\triangle ECA$, then form a proportion from corresponding sides.

**34.** $2402.2 \text{ cm}^2$

### PART IV

**35.** See the detailed solution for the proof.

**36.** Depth = 8.5 ft, volume = $3{,}752 \text{ ft}^3$, time = 41 hours

In **PARTS II–IV**, you are required to show how you arrived at your answers. For sample methods of solutions, see the *Answers Explained* section.

# Answers Explained

## PART I

1. The only choice that does not have a triangular cross-section is the cylinder. The triangular cross-sections of the other three choices are shown in the accompanying figure.

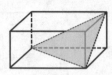

The correct choice is (**2**).

2. Reflections are rigid motions that always result in a congruent image. Dilations maintain the original shape but may change the size if the scale factor is different from 1. A dilation with a scale factor of $\frac{3}{2}$ will result in an image larger than the original.

The correct choice is (**4**).

3. Use the distance formula and the given coordinates $(-1, 5)$ and $(-3, 1)$ to find the length of side $\overline{RS}$.

$$
\begin{aligned}
d &= \sqrt{\left(x_2 - x_1\right)^2 + \left(y_2 - y_1\right)^2} \\
&= \sqrt{\left(-3 - (-1)\right)^2 + \left(1 - 5\right)^2} \\
&= \sqrt{\left(-2\right)^2 + \left(-4\right)^2} \\
&= \sqrt{20}
\end{aligned}
$$

Since we know the figure is a square with four sides of equal length, we can calculate the perimeter by multiplying the side length by 4. The perimeter is $4\sqrt{20}$.

The correct choice is (**3**).

**4.** The arcs between parallel chords are congruent, therefore m$\overarc{AC}$ = m$\overarc{BD}$ and each can be represented by $x$. The diameter $\overline{AOB}$ intercepts a semicircle whose arcs sum to 180°.

$$m\overarc{AC} + m\overarc{CD} + m\overarc{BD} = 180$$
$$x + 130° + x = 180°$$
$$2x + 130° = 180°$$
$$2x = 50°$$
$$x = 25°$$

The correct choice is **(1)**.

**5.** $\angle A \cong \angle D$ because they are alternate interior angles formed by the parallel segments $\overline{AB}$ and $\overline{DE}$. Also, the vertical angles $\angle ACB$ and $\angle DCE$ are congruent. Therefore, $\triangle ACB$ is similar to $\triangle DCE$ by the AA theorem and corresponding sides lengths are proportional.

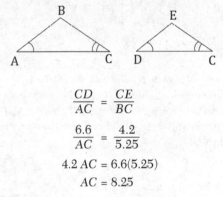

$$\frac{CD}{AC} = \frac{CE}{BC}$$
$$\frac{6.6}{AC} = \frac{4.2}{5.25}$$
$$4.2\,AC = 6.6(5.25)$$
$$AC = 8.25$$

The correct choice is **(4)**.

**6.** A reflection over the *x*-axis would map the top and bottom halves of the rectangle onto each other. The line *x* = 4 is a vertical line through the center of the rectangle, and it would map the left and right halves onto each other. The point (4, 0) is in the center of the rectangle. A rotation of 180° about that center point would also result in the rectangle mapping onto itself. The only choice that does not map the rectangle onto itself is a 180° rotation about the origin.

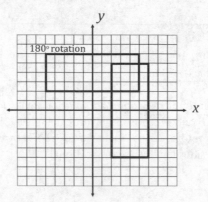

The correct choice is **(3)**.

**7.** Sketch the two triangles separately as shown to help identify congruent angles and corresponding sides. The parallel segments $\overline{BE}$ and $\overline{CD}$ form congruent alternate interior angles $\angle ABE$ and $\angle ACD$. $\angle A$ is a shared angle and is congruent to itself. Therefore, the two triangles are similar by the AA postulate. Write a proportion using corresponding sides to solve for the length *AE*.

$$\frac{AC}{AB} = \frac{AD}{AE}$$

$$\frac{4.5}{1} = \frac{18}{AE}$$

$$4.5\,AE = (1)(18)$$

$$AE = 4$$

The correct choice is **(4)**.

8.  Opposite angles of a parallelogram are congruent, so m$\angle R$ = m$\angle C$ = 70°.
    $\angle KSO$ is an exterior angle of $\triangle RSO$. From the exterior angle theorem
    we know the measure of the exterior angle is equal to the sum of the two
    further interior angles.

    $$m\angle KSO = m\angle R + m\angle ROS$$
    $$= 70° + 65°$$
    $$= 135°$$

    The correct choice is **(4)**.

9.  Check if the two pairs of corresponding sides are proportional. Form
    ratios using the two shorter sides and the two longer sides. If it is a true
    proportion the cross products will be equal.

    $$\frac{AR}{GR} = \frac{TR}{SR}$$
    $$\frac{15}{36} = \frac{18}{45}$$
    $$15 \cdot 45 = 18 \cdot 36$$
    $$675 \neq 638$$

    The cross products are not equal, so the two pairs of sides are not propor-
    tional. The triangles are not similar.

    The correct choice is **(4)**.

10. Dilating a line through the origin will multiply the $y$-intercept by the
    scale factor, but leave the slope unchanged. We need to find which
    choice has the same slope as the given line. Rewrite the given equation in
    $y = mx + b$ form to identify the slope.

    $$4y = 3x + 7$$
    $$y = \frac{3}{4}x + \frac{7}{4}$$

    The $y$-intercept is $\frac{7}{4}$ and the slope is $\frac{3}{4}$. Now check the slope of each
    choice by writing in $y = mx + b$ form. For choice (1) we have

    $$3x - 4y = 9$$
    $$-4y = -3x + 9$$
    $$y = \frac{3}{4}x - \frac{9}{4}$$

    Equation (1) also has a slope of $\frac{3}{4}$, so this must be the dilation.

    The correct choice is **(1)**.

**11.** The isosceles triangle theorem states that if $\overline{AB} \cong \overline{BC}$ then the opposite angles must also be congruent. Therefore, $\angle BAC \cong \angle BCA$ and we can represent both angles with the variable $y$ (not $x$ since that variable is already used for another angle!). Apply the angle sum theorem in $\triangle ABC$.

$$\text{m}\angle A + \text{m}\angle B + \text{m}\angle ACB = 180°$$
$$y + 62° + y = 180°$$
$$2y + 62° = 180°$$
$$2y = 118°$$
$$y = 59°$$

$\text{m}\angle A$ and $\text{m}\angle ACB$ both measure 59°. The angle labeled $x$ is an exterior angle, so its measure is equal to the sum of the two furthest interior angles.

$$x = \text{m}\angle A + \text{m}\angle B$$
$$x = 59° + 62°$$
$$x = 121°$$

The correct choice is **(4)**.

**12.** The lengths of the secant and tangent are related by the formula

$$(\text{secant})^2 = (\text{outside part of tangent}) \cdot (\text{entire tangent})$$

$$PA^2 = PB(PB + PC)$$
$$PA^2 = 3(3 + 15)$$
$$PA^2 = 3 \cdot 18$$
$$PA^2 = 54$$
$$PA = \sqrt{54}$$

Since this is not one of the choices we simplify the radical. Look for a factor 54 that is a perfect square. The desired factors are 9 and 6, where 9 is the perfect square.

$$\sqrt{54} = \sqrt{9 \cdot 6}$$
$$= 3\sqrt{6}$$

The correct choice is **(2)**.

As an alternative to simplifying the radical, you can type the radical into your calculator to find the decimal approximation 7.34846. Check each of the choices to find the one with the matching decimal approximation.

**13.** Rotating a rectangle around a side will form a cylinder. The length of the side it is rotated around is the height and the other side gives the radius as shown in figures (a) and (b) below. The situation is similar if the rectangle is rotated about a line of symmetry, except that one of the sides gives the diameter instead of the radius as shown in figure (c) below.

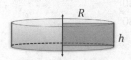  

(a) rotation about　　　(b) rotation about　　　(c) rotation about the
   the short side　　　   the long side　　　   line of symmetry

Calculate the volume for each choice using the volume formula $V = \pi R^2 h$, which is provided on your reference sheet.

**Choice 1:**

$h = 10, R = 6$

$V = \pi R^2 h$

$\quad = \pi \cdot 6^2 \cdot 10$

$\quad = 360\pi$

**Choice 2:**

$h = 6, R = 10$

$V = \pi R^2 h$

$\quad = \pi \cdot 10^2 \cdot 6$

$\quad = 600\pi$

**Choice 3:**

$h = 6, R = 5$

$V = \pi R^2 h$

$\quad = \pi \cdot 5^2 \cdot 6$

$\quad = 150\pi$

**Choice 4:**

$h = 10, R = 3$

$V = \pi R^2 h$

$\quad = \pi \cdot 3^2 \cdot 10$

$\quad = 90\pi$

The correct choice is **(3)**.

**14.** A parallelogram whose diagonals are perpendicular is a rhombus. $\overline{AC} \perp \overline{BD}$ represents the two perpendicular diagonals.

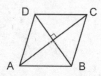

The correct choice is **(3)**.

The other possible rhombus properties that could be used are consecutive sides congruent or diagonals bisect the vertex angles. However, none of the choices represents these properties.

**15.** Find the unknown angle by setting up a trig ratio using an inverse trig function. Relative to the angle $x$, the opposite side has length 1 and the adjacent side has length 12. These suggest using the tangent.

$$\tan(x) = \frac{\text{opposite}}{\text{adjacent}}$$

$$= \frac{1}{12}$$

$$x = \tan^{-1}\left(\frac{1}{12}\right)$$

$$= 4.76$$

The correct choice is **(1)**.

**16.** The three sides of $\triangle DEF$ are all midsegments because they are formed by joining midpoints of the sides of $\triangle ABC$. A midsegment has a length equal to $\frac{1}{2}$ of the opposite side, making every side in $\triangle DEF$ half the length of a corresponding side in $\triangle ABC$. Therefore, $\triangle DEF$ is the image after a dilation with a scale factor of $\frac{1}{2}$. Areas of similar figures are proportional to the scale factor squared.

$$\text{Ratio of areas} = \left(\frac{1}{2}\right)^2$$

$$= \frac{1}{4}$$

The correct choice is **(4)**.

**17.** Find the $x$- and $y$-coordinates of $P$ with the formula

$$\text{ratio} = \frac{x - x_1}{x_2 - x} \qquad \text{ratio} = \frac{y - y_1}{y_2 - y}$$

Using the coordinates $A(-8, -2)$ and $B(16, 6)$, we have $x_1 = -8$, $y_1 = -2$, $x_2 = 16$, and $y_2 = 6$. The ratio is $\frac{3}{5}$.

Solve for the $x$-coordinate:

$$\text{ratio} = \frac{x - x_1}{x_2 - x}$$

$$\frac{3}{5} = \frac{x - (-8)}{16 - x}$$

$$5(x + 8) = 3(16 - x) \qquad \text{cross-multiply}$$

$$5x + 40 = 48 - 3x$$

$$8x = 8$$

$$x = 1$$

Solve for the $y$-coordinate:

$$\text{ratio} = \frac{y - y_1}{y_2 - y}$$

$$\frac{3}{5} = \frac{y - (-2)}{6 - y}$$

$$5(y + 2) = 3(6 - y) \qquad \text{cross-multiply}$$

$$5y + 10 = 18 - 3y$$

$$8y = 8$$

$$y = 1$$

The coordinates of $P$ are $(1, 1)$.

The correct choice is (**1**).

**18.** One possible approach for this problem is to calculate the measures of ∠K and ∠M using the given lengths and assuming ∠KLM is a right angle. If the sum of ∠K and ∠M is 90°, then our assumption was correct and ∠KLM must be a right angle. Using choice (2), we find

$$\cos M = \left(\frac{9}{12}\right) \text{ in } \triangle MNL \qquad\qquad \sin K = \frac{12}{16} \text{ in } \triangle KLM$$

$$M = \cos^{-1}\frac{9}{12} \qquad\qquad K = \sin^{-1}\frac{12}{16}$$

$$= 41.4° \qquad\qquad\qquad = 48.6°$$

Since 41.4° + 48.6° = 90°, we know the third angle is 90° as well.

The correct choice is **(2)**.

**19.** Apply a trig ratio to find *BC*. Relative to ∠A, $\overline{AC}$ is the hypotenuse and $\overline{BC}$ is the opposite. We use the sine ratio.

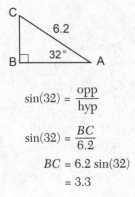

$$\sin(32) = \frac{\text{opp}}{\text{hyp}}$$

$$\sin(32) = \frac{BC}{6.2}$$

$$BC = 6.2\sin(32)$$

$$= 3.3$$

The correct choice is **(1)**.

**20.** The population density, area, and population are related by the formula

$$\text{population density} = \frac{\text{population}}{\text{area}}$$

Since we need to find the area of four states, it will be easiest to solve for area in terms of the other variables.

$$\text{population density} = \frac{\text{population}}{\text{area}}$$

$$\text{area} \cdot \text{population density} = \text{population}$$

$$\text{area} = \frac{\text{population}}{\text{population density}}$$

**Florida**

$$\text{area} = \frac{18{,}801{,}310}{350.6}$$

$$= 53{,}626 \text{ miles}^2$$

**New York**

$$\text{area} = \frac{19{,}378{,}102}{411.2}$$

$$= 47{,}126 \text{ miles}^2$$

**Illinois**

$$\text{area} = \frac{12{,}830{,}632}{231.1}$$

$$= 55{,}520$$

**Pennsylvania**

$$\text{area} = \frac{12{,}702{,}379}{283.9}$$

$$= 44{,}742$$

The order of states from largest area to smallest is Illinois, Florida, New York, and Pennsylvania.

The correct choice is (**1**).

**21.** The co-function relationship states that if a sine and cosine are equal, then the angles must sum to 90°. Set the sum of the two angles to 90° and then solve for $x$.

$$40 - x + 3x = 90$$
$$40 + 2x = 90$$
$$2x = 50$$
$$x = 25$$

The correct choice is (**4**).

**22.** A rotation of any multiple of $\dfrac{360}{n}$, where $n$ is the number of sides, will map a regular polygon onto itself. A decagon has ten sides, so the angle must be a multiple of $\dfrac{360}{10}$ or 36°. Divide each choice by 36°. The one with an integer quotient is a multiple of 36°.

$$\frac{10}{36} = 0.27 \qquad\qquad \frac{225}{36} = 6.25$$

$$\frac{150}{36} = 4.17 \qquad\qquad \frac{252}{36} = 7$$

A rotation of 252° will map a decagon onto itself.

The correct choice is **(4)**.

**23.** The radius of the circle is half the diameter, or 16. Apply the formula for the area of a sector to find the angle.

$$A_{\text{sector}} = \frac{1}{2} R^2, \text{ where } \theta \text{ is the central angle measured in radians}$$

$$\frac{512\pi}{3} = \frac{1}{2} \cdot 16^2$$

$$\frac{1024\pi}{3} = 16^2$$

$$= \frac{1024\pi}{3 \cdot 16^2}$$

$$= \frac{4\pi}{3}$$

The correct choice is **(2)**.

**24.** The perpendicular bisector passes through the midpoint of the given segment, and has a slope equal to the negative reciprocal of the segment's slope. Find the midpoint using the endpoints $(-2, 4)$ and $(2, -4)$ and the midpoint formula.

$$\text{midpoint} = \left( \frac{x_1 + x_2}{2}, \frac{y_1 + y_2}{2} \right)$$

$$= \left( \frac{-2 + 2}{2}, \frac{4 + (-4)}{2} \right)$$

$$= (0, 0)$$

Next find the slope of the segment.

$$\text{slope} = \frac{y_2 - y_1}{x_2 - x_1}$$

$$= \frac{-4 - 4}{2 - (-2)}$$

$$= \frac{-8}{4}$$

$$= -2$$

The negative reciprocal of the slope is $\frac{1}{2}$. Using the point–slope form of a line with a slope of $\frac{1}{2}$ and the point $(0, 0)$ we have

$$y - y_1 = m(x - x_1)$$

$$y - 0 = \frac{1}{2}(x - 0)$$

$$y = \frac{1}{2}x$$

$$y - \frac{1}{2}x = 0$$

$$2y - x = 0 \qquad \text{multiply both sides by 2 to eliminate the fraction}$$

The correct choice is **(4)**.

## PART II

**25.** The volumes are the same. Each cylinder has a radius of 5 meters, so the cross-sectional areas are equal. Also they each have a height of 11.5 meters. Cavalieri's principle states that if two solids have equal cross-sectional areas and heights, then their volumes are equal. *Note that calculating the volume of each cylinder would earn only 1 of the 2 possible credits.*

**26.** The diagonals of a rhombus, like all parallelograms, bisect each other. Calculate the lengths of $\overline{TS}$ and $\overline{TR}$ by dividing each diagonal by 2. $TS = 15$ and $TR = 8$. The diagonals of rhombuses are also perpendicular, so $\triangle STR$ is a right triangle. $SR$ can be found using the Pythagorean theorem.

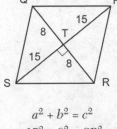

$$a^2 + b^2 = c^2$$
$$15^2 + 8^2 = SR^2$$
$$289 = SR^2$$
$$SR = 17$$

All four sides of a rhombus are congruent, so the perimeter is equal to four times the side length.

$$\text{perimeter} = 4 \cdot 17$$
$$= \mathbf{68}$$

**27.** A rotation of 180° about the origin followed by a translation of 1 unit up and 1 unit left will map *MATH* to *M"A"T"H"*.

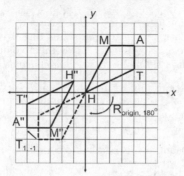

**28.** Mark point *A* anywhere on the circle. Using the same radius as the circle and the point of the compass at *A*, make an arc intersecting the circle at *B*. With the same compass opening, place the point of the compass at *B* and make an arc at *C*. Continue making arcs intersecting at *D*, *E*, and *F*. *ABCDEF* is a regular hexagon.

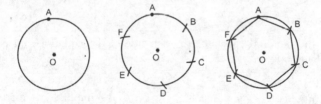

**29.** The length of $\overline{AB}$ can be calculated using the distance formula and coordinates $A(2, 3)$ and $B(5, -1)$.

$$d = \sqrt{(x_2 - x_1)^2 + (y_2 - y_1)^2}$$

$$= \sqrt{(5-2)^2 + (-1-3)^2}$$

$$= \sqrt{25}$$

$$= 5$$

A dilation of $\frac{1}{2}$ will multiply the length of the segment by $\frac{1}{2}$, so the length of the image is $\frac{5}{2}$.

**30.** Rigid motions preserve angle measure and length, so $\angle A \cong \angle X$, $\angle C \cong \angle Z$, and $\overline{AC} \cong \overline{XZ}$. The triangles are therefore congruent by the ASA postulate. $\overline{BC} \cong \overline{YZ}$ because corresponding parts of congruent triangles are congruent.

**31.** Use completing the square to rewrite the equation in the form $(x - h)^2 + (y - k)^2 = r^2$. First group the $x$-terms and the $y$-terms on the left, and move the constant term to the right.

$$x^2 + y^2 - 6x = 56 - 8y$$
$$x^2 - 6x + y^2 + 8y = 56$$

Next find the constant terms needed to complete the square:

| constant for the $x$-terms | constant for the $y$-terms |
|---|---|
| $\left(\dfrac{1}{2} \text{ coefficient of } x\right)^2$ | $\left(\dfrac{1}{2} \text{ coefficient of } y\right)^2$ |
| $= \left(\dfrac{1}{2} \cdot (-6)\right)^2$ | $= \left(\dfrac{1}{2} \cdot 8\right)^2$ |
| $= (-3)^2$ | $= (4)^2$ |
| $= 9$ | $= 16$ |

Add the required constants to each side of the equation and factor the left-hand side.

$$x^2 - 6x + 9 + y^2 + 8y + 16 = 56 + 9 + 16 \quad \text{add 9 and 16 to each side}$$
$$x^2 - 6x + 9 + y^2 + 8y + 16 = 81 \quad \text{simplify}$$
$$(x - 3)^2 + (y + 4)^2 = 9 \quad \text{factor}$$

The values of $h$ and $k$ are 3 and $-4$. Be careful with the signs because $h$ and $k$ are the values subtracted from $x$ and $y$. The center is located at $(3, -4)$. Since $r^2 = 9$, the radius is equal to 3.

center: **(3, −4)**     radius: **(3)**.

## PART III

**32.** Find the coordinates of the midpoints $A$ and $B$ using the midpoint formula.

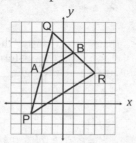

midpoint of $P(-3, -1)$ $Q(-1, 7)$ 　　　midpoint of $Q(-1, 7)$ $R(3, 3)$

midpoint $= \left( \dfrac{x_1 + x_2}{2}, \dfrac{y_1 + y_2}{2} \right)$ 　　　midpoint $= \left( \dfrac{x_1 + x_2}{2}, \dfrac{y_1 + y_2}{2} \right)$

$\quad = \left( \dfrac{-3 + (-1)}{2}, \dfrac{-1 + 7}{2} \right)$ 　　　$\quad = \left( \dfrac{-1 + 3}{2}, \dfrac{7 + 3}{2} \right)$

$\quad = (-2, 3)$ 　　　$\quad = (1, 5)$

Next find the length and slope of segments $\overline{AB}$ and $\overline{PR}$.

slope of $\overline{AB} = \dfrac{y_2 - y_1}{x_2 - x_1}$ 　　　slope of $\overline{PR} = \dfrac{y_2 - y_1}{x_2 - x_1}$

$\quad = \dfrac{5 - 3}{1 - (-2)}$ 　　　$\quad = \dfrac{3 - (-1)}{3 - (-3)}$

$\quad = \dfrac{2}{3}$ 　　　$\quad = \dfrac{4}{6}$

　　　　　　　　　　　　$\quad = \dfrac{2}{3}$

$$AB = \sqrt{(x_2 - x_1)^2 + (y_2 - y_1)^2} \qquad PR = \sqrt{(x_2 - x_1)^2 + (y_2 - y_1)^2}$$

$$= \sqrt{(1-(-2))^2 + (5-3)^2} \qquad\qquad = \sqrt{(3-(-3))^2 + (3-(-1))^2}$$

$$= \sqrt{9+4} \qquad\qquad\qquad\qquad\quad = \sqrt{36+16}$$

$$= \sqrt{13} \qquad\qquad\qquad\qquad\quad = \sqrt{52}$$

$$\qquad\qquad\qquad\qquad\qquad\quad = \sqrt{4\cdot 13}$$

$$\qquad\qquad\qquad\qquad\qquad\quad = 2\sqrt{13}$$

$\overline{AB}$ is parallel to $\overline{PR}$ because they have the same slope. $\overline{AB}$ is half the length of $\overline{PR}$ because $\sqrt{13}$ is one-half of $2\sqrt{13}$.

33. To prove two of the pairs of sides are proportional we can prove $\triangle ABC$ is similar to $\triangle ECA$. $\angle B$ is a right angle because it is an inscribed angle that intercepts a diameter. $\angle ECA$ is a right angle because it is formed by a tangent and radius. Therefore, $\angle B \cong \angle ECA$. We also have $\angle BCA \cong \angle ECA$ because they are the alternate interior angles formed by the given parallel lines. Since two pairs of angles are congruent, $\triangle ABC$ is similar to $\triangle ECA$ by the AA postulate. $\dfrac{BC}{CA} = \dfrac{AB}{EC}$ because the corresponding parts of similar triangles are proportional.

34. Let $s$ represent the side length of the poster. The 58 centimeter diagonal forms a right isosceles triangle with side length $s$. Use the Pythagorean theorem to solve for $s$.

$$a^2 + b^2 = c^2$$
$$s^2 + s^2 = 58^2$$
$$2s^2 = 3{,}364$$
$$s^2 = 1{,}682$$
$$s^2 = \sqrt{1{,}682}$$
$$s = 41.012$$

The length of the frame $L$ can be found by adding 4 centimeters to each side of the poster.

$$L = 41.012 + 4 + 4$$
$$= \mathbf{49.012}$$

The area of the square frame is found using the area formula for a square.

$$\text{Area} = L^2$$
$$= 49.012^2$$
$$= 2402.2195$$
$$= \mathbf{2402.2 \ cm^2}$$

## PART III

**35.** The strategy for this proof is to show △DEC is isosceles. The congruent legs $\overline{DE}$ and $\overline{EC}$ can be used to prove △ADE and △BCE are congruent. CPCTC can then be used to show △AEB is isosceles.

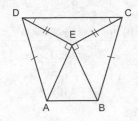

| Statement | Reason |
|---|---|
| 1. ∠CDE ≅ ∠DEC | 1. Given |
| 2. $\overline{DE} \cong \overline{EC}$ | 2. Sides of a triangle opposite congruent angles are congruent |
| 3. $\overline{AE}$ and $\overline{DE}$, $\overline{BE}$ and $\overline{CE}$ | 3. Given |
| 4. AED and BEC are right angles | 4. Perpendicular lines form right angles |
| 5. △AED and △BEC are right triangles | 5. A triangle with a right angle is a right triangle |
| 6. Isosceles trapezoid ABCD | 6. Given |
| 7. $\overline{AD} \cong \overline{BC}$ | 7. Legs of an isosceles trapezoid are congruent |
| 8. △AED ≅ △BEC | 8. HL |
| 9. $\overline{AE} \cong \overline{BE}$ | 9. CPCTC |
| 10. △AEB is isosceles | 10. A triangle with two congruent sides is isosceles |

**36.** The pool is composed of three rectangular prisms and a triangular prism as shown in the figure below. The dimensions $x$ and $y$ are needed in the triangular prism in order to calculate its base area. The dimension $y$ is also needed to determine the height of prism D.

$$x = 35 - 9 - 12.5$$
$$= 13.5$$

Apply the tangent ratio to find the dimension $x$.

$$\tan(16.5) = \frac{y}{13.5}$$
$$y = 13.5 \tan(16.5)$$
$$= 3.9988$$

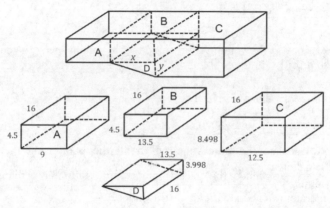

Now calculate the volume of each prism.

**Prism A**

$V = Bh$

$\quad = (9 \cdot 16)4.5$

$\quad = 648 \text{ ft}^3$

**Prism C**

$V = Bh$

$\quad = \left(\frac{1}{2}13.5 \cdot 3.9988\right)16$

$\quad = 431.8704 \text{ ft}^3$

**Prism B**

$V = Bh$

$\quad = (13.5 \cdot 16)4.5$

$\quad = 972 \text{ ft}^3$

**Prism D**

$V = Bh$

$\quad = (12.5 \cdot 16)(4.5 + 3.9988)$

$\quad = 1699.76 \text{ ft}^3$

The total volume of the pool is

$$648 + 972 + 431.8704 + 1699.76 = \mathbf{3751.6304}.$$

Rounded to the nearest cubic foot, the volume is 3752 ft$^3$.

To find how long it will take to fill the pool 6 inches from the top we need to subtract the volume of a thin 6-inch-high rectangular prism from the total volume. This volume is taken from the very top of the pool, so its dimensions are 16 feet × 35 feet × 6 inches. Convert the 6 inches to feet before calculating the volume.

$$6 \text{ inches} \cdot \frac{1 \text{ foot}}{12 \text{ inches}} = 0.5 \text{ foot}$$

The volume of the top layer = $B\,h$

$$= (16 \cdot 35)\,0.5$$

$$= 280 \text{ ft}^3$$

The volume to be filled = 3752 ft$^3$ – 280 ft$^3$

$$= 3472 \text{ ft}^3$$

Next, convert the volume from cubic feet to gallons using the conversion ratio provided in the problem.

$$\text{Number of gallons} = 3472 \text{ ft}^3 \cdot \frac{7.48 \text{ gallons}}{\text{ft}^3}$$

$$= 25970.56 \text{ gallons}$$

Now calculate the time required to fill the pool using the fill rate $\frac{10.5 \text{ gallons}}{\text{minute}}$. Since the gallons per minute ratio has gallons in the numerator, we need to *divide* by this ratio.

$$\text{time} = 25970.56 \text{ gallons} \div \frac{10.5 \text{ gallons}}{\text{minute}}$$

$$= 25970.56 \cdot \frac{1 \text{ minute}}{10.5 \text{ gallons}}$$

$$= 2473.38 \text{ minutes}$$

Finally, convert the time in minutes to time in hours using the ratio $\frac{1 \text{ hour}}{60 \text{ minutes}}$

$$\text{time} = 2473.38 \text{ minutes} \cdot \frac{1 \text{ hour}}{60 \text{ minutes}}$$

$$= 41.22 \text{ hours}$$

Rounded to the nearest hour, it will take 41 hours to fill the pool.

# HOW TO CONVERT YOUR RAW SCORE TO YOUR GEOMETRY REGENTS EXAMINATION SCORE

The conversion chart below can be used to determine your final score on the August 2017 Regents Exam in Geometry. To find your final exam score, locate in the "Raw Score" column the total number of points you scored out of a possible 86. Then locate in the adjacent column to the right the scale score that corresponds to your raw score. The scale score is your final score.

| Raw Score | Scale Score | Performance Level | Raw Score | Scale Score | Performance Level | Raw Score | Scale Score | Performance Level |
|---|---|---|---|---|---|---|---|---|
| 86 | 100 | 5 | 57 | 79 | 3 | 28 | 59 | 2 |
| 85 | 99 | 5 | 56 | 79 | 3 | 27 | 58 | 2 |
| 84 | 97 | 5 | 55 | 78 | 3 | 26 | 57 | 2 |
| 83 | 96 | 5 | 54 | 78 | 3 | 25 | 55 | 2 |
| 82 | 95 | 5 | 53 | 77 | 3 | 24 | 54 | 1 |
| 81 | 94 | 5 | 52 | 77 | 3 | 23 | 53 | 1 |
| 80 | 93 | 5 | 51 | 76 | 3 | 22 | 51 | 1 |
| 79 | 92 | 5 | 50 | 76 | 3 | 21 | 50 | 1 |
| 78 | 91 | 5 | 49 | 75 | 3 | 20 | 48 | 1 |
| 77 | 90 | 5 | 48 | 75 | 3 | 19 | 47 | 1 |
| 76 | 90 | 5 | 47 | 74 | 3 | 18 | 45 | 1 |
| 75 | 89 | 5 | 46 | 74 | 3 | 17 | 43 | 1 |
| 74 | 88 | 5 | 45 | 73 | 3 | 16 | 42 | 1 |
| 73 | 87 | 5 | 44 | 73 | 3 | 15 | 40 | 1 |
| 72 | 87 | 5 | 43 | 72 | 3 | 14 | 38 | 1 |
| 71 | 86 | 5 | 42 | 72 | 3 | 13 | 36 | 1 |
| 70 | 86 | 5 | 41 | 71 | 3 | 12 | 34 | 1 |
| 69 | 86 | 5 | 40 | 70 | 3 | 11 | 32 | 1 |
| 68 | 85 | 5 | 39 | 70 | 3 | 10 | 30 | 1 |
| 67 | 84 | 4 | 38 | 69 | 3 | 9 | 28 | 1 |
| 66 | 83 | 4 | 37 | 68 | 3 | 8 | 25 | 1 |
| 65 | 83 | 4 | 36 | 67 | 3 | 7 | 23 | 1 |
| 64 | 82 | 4 | 35 | 66 | 3 | 6 | 20 | 1 |
| 63 | 82 | 4 | 34 | 66 | 3 | 5 | 18 | 1 |
| 62 | 82 | 4 | 33 | 65 | 3 | 4 | 15 | 1 |
| 61 | 81 | 4 | 32 | 64 | 2 | 3 | 12 | 1 |
| 60 | 81 | 4 | 31 | 63 | 2 | 2 | 8 | 1 |
| 59 | 80 | 4 | 30 | 61 | 2 | 1 | 5 | 1 |
| 58 | 80 | 3 | 29 | 60 | 2 | 0 | 0 | 1 |

# Appendix

## THE COMMON CORE GEOMETRY LEARNING STANDARDS

The New York State Geometry Regents is aligned to the Geometry Common Core Learning Standards. All topics listed below may be found on the Regents exam.

### CONGRUENCE G-CO

**Experiment with Transformations in the Plane**

1. Know precise definitions of angle, circle, perpendicular line, parallel line, and line segment, based on the undefined notions of point, line, distance along a line, and distance around a circular arc.

2. Represent transformations in the plane using, for example, transparencies and geometry software; describe transformations as functions that take points in the plane as inputs and give other points as outputs. Compare transformations that preserve distance and angle to those that do not (e.g., translation versus horizontal stretch).

3. Given a rectangle, parallelogram, trapezoid, or regular polygon, describe the rotations and reflections that carry it onto itself.

4. Develop definitions of rotations, reflections, and translations in terms of angles, circles, perpendicular lines, parallel lines, and line segments.

5. Given a geometric figure and a rotation, reflection, or translation, draw the transformed figure using, for example, graph paper, tracing paper, or geometry software. Specify a sequence of transformations that will carry a given figure onto another.

## Understand Congruence in Terms of Rigid Motions

6. Use geometric descriptions of rigid motions to transform figures and to predict the effect of a given rigid motion on a given figure; given two figures, use the definition of congruence in terms of rigid motions to decide if they are congruent.

7. Use the definition of congruence in terms of rigid motions to show that two triangles are congruent if and only if corresponding pairs of sides and corresponding pairs of angles are congruent.

8. Explain how the criteria for triangle congruence (ASA, SAS, and SSS) follow from the definition of congruence in terms of rigid motions.

## Prove Geometric Theorems

9. Prove theorems about lines and angles. *Theorems include: vertical angles are congruent; when a transversal crosses parallel lines, alternate interior angles are congruent and corresponding angles are congruent; points on a perpendicular bisector of a line segment are exactly those equidistant from the segment's endpoints.*

10. Prove theorems about triangles. *Theorems include: measures of interior angles of a triangle sum to 180°; base angles of isosceles triangles are congruent; the segment joining midpoints of two sides of a triangle is parallel to the third side and half the length; the medians of a triangle meet at a point.*

11. Prove theorems about parallelograms. *Theorems include: opposite sides are congruent; opposite angles are congruent; the diagonals of a parallelogram bisect each other; and conversely, rectangles are parallelograms with congruent diagonals.*

## Make Geometric Constructions

12. Make formal geometric constructions with a variety of tools and methods (compass and straightedge, string, reflective devices, paper folding, dynamic geometric software, etc.). *Copying a segment; copying an angle; bisecting a segment; bisecting an angle; constructing perpendicular lines, including the perpendicular bisector of a line segment; and constructing a line parallel to a given line through a point not on the line.*

13. Construct an equilateral triangle, a square, and a regular hexagon inscribed in a circle.

# SIMILARITY, RIGHT TRIANGLES, AND TRIGONOMETRY G-SRT

## Understand Similarity in Terms of Similarity Transformations

1. Verify experimentally the properties of dilations given by a center and a scale factor:

   a. A dilation takes a line not passing through the center of the dilation to a parallel line and leaves a line passing through the center unchanged.

   b. The dilation of a line segment is longer or shorter in the ratio given by the scale factor.

2. Given two figures, use the definition of similarity in terms of similarity transformations to decide if they are similar; explain using similarity transformations the meaning of similarity for triangles as the equality of all corresponding pairs of angles and the proportionality of all corresponding pairs of sides.

3. Use the properties of similarity transformations to establish the AA criterion for two triangles to be similar.

## Prove Theorems Involving Similarity

4. Prove theorems about triangles. *Theorems include: a line parallel to one side of a triangle divides the other two proportionally and conversely; the Pythagorean theorem proved using triangle similarity.*

5. Use congruence and similarity criteria for triangles to solve problems and to prove relationships in geometric figures.

## Define Trigonometric Ratios and Solve Problems Involving Right Triangles

6. Understand that by similarity, side ratios in right triangles are properties of the angles in the triangle, leading to definitions of trigonometric ratios for acute angles.

7. Explain and use the relationship between the sine and cosine of complementary angles.

8. Use trigonometric ratios and the Pythagorean theorem to solve right triangles in applied problems.

# CIRCLES G-C

**Understand and Apply Theorems About Circles**

1. Prove that all circles are similar.
2. Identify and describe relationships among inscribed angles, radii, and chords. *Include the relationship between central, inscribed, and circumscribed angles; inscribed angles on a diameter as right angles; the radius of a circle being perpendicular to the tangent where the radius intersects the circle.*
3. Construct the inscribed and circumscribed circles of a triangle, and prove properties of angles for a quadrilateral inscribed in a circle.

**Find Arc Lengths and Areas of Sectors of Circles**

4. Derive using similarity the fact that the length of the arc intercepted by an angle is proportional to the radius, and define the radian measure of the angle as the constant of proportionality; derive the formula for the area of a sector.

# EXPRESSING GEOMETRIC PROPERTIES WITH EQUATIONS G-GPE

**Translate Between the Geometric Description and the Equation for a Conic Section**

1. Derive the equation of a circle of given center and radius using the Pythagorean theorem; complete the square to find the center and radius of a circle given by an equation.
2. Derive the equation of a parabola given a focus and directrix.

**Use Coordinates to Prove Simple Geometric Theorems Algebraically**

3. Use coordinates to prove simple geometric theorems algebraically. *For example, prove or disprove that a figure defined by four given points in the coordinate plane is a rectangle; prove or disprove that the point $(1, \sqrt{3})$ lies on the circle centered at the origin and containing the point $(0, 2)$.*
4. Prove the slope criteria for parallel and perpendicular lines and use them to solve geometric problems (e.g., find the equation

of a line parallel or perpendicular to a given line that passes through a given point).

5. Find the point on a directed line segment between two given points that partitions the segment in a given ratio.
6. Use coordinates to compute perimeters of polygons and areas of triangles and rectangles (e.g., using the distance formula).

## GEOMETRIC MEASUREMENT AND DIMENSION G-GMD

### Explain Volume Formulas and Use Them to Solve Problems

1. Give an informal argument for the formulas for the circumference of a circle, area of a circle, volume of a cylinder, pyramid, and cone. *Use dissection arguments, Cavalieri's principle, and informal limit arguments.*
2. Use volume formulas for cylinders, pyramids, cones, and spheres to solve problems.

### Visualize Relationships Between Two-dimensional and Three-dimensional Objects

3. Identify the shapes of 2-dimensional cross-sections of 3-dimensional objects, and identify 3-dimensional objects generated by rotations of 2-dimensional objects.

## MODELING WITH GEOMETRY G-MG

### Apply Geometric Concepts in Modeling Situations

1. Use geometric shapes, their measures, and their properties to describe objects (e.g., modeling a tree trunk or a human torso as a cylinder).
2. Apply concepts of density based on area and volume in modeling situations (e.g., persons per square mile, BTUs per cubic foot).
3. Apply geometric methods to solve design problems (e.g., designing an object or structure to satisfy physical constraints or minimize cost; working with typographic grid systems based on ratios).